America's Public Lands

America's Public Lands

From Yellowstone to
Smokey Bear and Beyond

Randall K. Wilson

ROWMAN & LITTLEFIELD
Lanham • Boulder • New York • Toronto • Plymouth, UK

Published by Rowman & Littlefield
4501 Forbes Boulevard, Suite 200, Lanham, Maryland 20706
www.rowman.com

10 Thornbury Road, Plymouth PL6 7PP, United Kingdom

British Library Cataloguing in Publication Information Available

Library of Congress Cataloging-in-Publication Data
Wilson, Randall K., 1966–
  America's public lands : from Yellowstone to Smokey Bear and beyond / Randall K. Wilson.
    pages cm.
  Includes bibliographical references and index.
  ISBN 978-1-4422-0797-4 (cloth : alkaline paper) — ISBN 978-1-4422-0799-8 (electronic) 1. Public lands—United States. 2. Public lands—Government policy—United States. 3. Public lands—United States—Management—History. 4. Conservation of natural resources—United States. I. Title.
  HD216.W48 2014
  333.10973—dc23

                                                          2014000965

♾™ The paper used in this publication meets the minimum requirements of American National Standard for Information Sciences—Permanence of Paper for Printed Library Materials, ANSI/NISO Z39.48-1992.

Printed in the United States of America

For Robin, Orrin, Nolan, and Olivia Suzanne

Climb the mountains and get their good tidings, Nature's peace will flow into you as sunshine flows into trees. The winds will blow their own freshness into you, and the storms their energy, while cares will drop off like autumn leaves.

—John Muir, *Our National Parks*, 56

Contents

Acknowledgments

I could not have completed this book without the generous help of many people. At Gettysburg College, where I serve as a faculty member in the Department of Environmental Studies, I have been the beneficiary of an exceptionally talented group of students who have assisted me in this project, including Sara Cawley, Max Barresi, Will Boone, Michael Catalano, Emily Ruhl, Katie McCrea, and Kelly Webster. I am especially grateful to Jessica Lee, who lent her formidable cartographic skills to the task of designing the maps in this book.

In addition, I've learned a great deal from the discussions and constructive criticism offered by colleagues from a variety of disciplines, all of whom have contributed to my understanding of public land issues over the years, and some of whom were kind enough to provide critical feedback on specific chapters. I owe significant thanks in this regard to Thomas Crawford, Marla Emery, Patrick Hurley, Paul Robbins, David Reynolds, Rebecca Roberts, Claire Pavlik, Malcolm Rohrbough, Tim Shannon, Jay Turner, and Andrew Wilson. I am also grateful for the constructive comments offered by anonymous reviewers.

A number of individuals provided invaluable assistance in tracking down materials and images used in this book. I am indebted to James Lewis and Eben Lehman from the Forest History Society, Laura O'Leary from the Aldo Leopold Foundation, Lisa Dare from the Wilderness Society, Kristin Raveling, Beth Ullenberg, and Loir Iverson from the U.S. Fish and Wildlife Service, Kristin Ellis from the Ad Council, Stacy Mason from the National Park Service, and Courtney Whiteman from the Bureau of Land Management.

For years I've lived, conducted research, and taught field courses in southwestern Colorado. I've profited greatly over the years from the insights, observations, and general wisdom offered by Dr. Sam Burns, former director of the Center for Community Services at Fort Lewis College, and from Mike Preston, former director of the Montezuma County Public Lands Office. I must also thank Tim Richard, Carla Garrison Harper, and Mary Hobbs. From the San Juan National Forest, I am deeply indebted to staff members (both former and current) Sharon Hatch, Phil Kemp, Jim Powers, Cal Joyner, Thurman Wilson, and Mike Znerold.

A significant portion of the writing was completed while living, teaching, and studying abroad in northern England and Austria. I am thankful to Gettysburg College and the U.S. Fulbright Commission for providing these unique teaching and research opportunities, including the valuable perspectives that sometimes only show themselves when one has the chance to take a step back from the subject (in my case, several thousand miles). I am also exceedingly grateful to Susan McEachern, both for getting this project started and helping me bring it to completion. I also must thank Carrie Broadwell-Tkach, Jehanne Schweitzer, Desiree Reid, and Jennifer Rushing-Schurr for their assistance and expertise in the publication process. Of course, any shortcomings or errors contained in the text are my own.

Finally, it is worth noting that my interest in America's public lands goes well beyond academics, stretching back to the camping and backpacking trips of my youth. Led by my brother, Doug, a few close friends and I spent memorable weeks in the Sierra Nevadas each summer. Later on, attending college at Humboldt State in California's redwood country, combined with years spent wrangling on a Colorado dude ranch, solidified a lifelong passion for the wild and open spaces that serve as the subject of this book. For the past fourteen years, I've lived as a "transplanted westerner" in southern Pennsylvania. But during this time, I've gained a deep appreciation for the truly national scope of America's public land system. Exploring the landscapes and historic sites of the Mid-Atlantic and the eastern seaboard is something my family and I have come to thoroughly enjoy. As always, it is to my family that I owe the most. Robin, Orrin, Nolan, and Olivia have put up with numerous absences on my part—both mental and physical—during the long hours I've devoted to this project. It is definitely time for a hike!

Introduction
Why Public Lands?

Whether we realize it or not, every citizen of the United States is a landowner. Collectively, we are heir to approximately 650 million acres of federally managed public lands, nearly one-third of the entire U.S. land area. Yet when we think of public lands, often the first images that come to mind are those associated with the national parks: Yellowstone's Old Faithful, the Grand Canyon, or Half Dome in Yosemite Valley. Unless one resides in the rural West or is otherwise involved with public lands issues, the 247 million acres managed by the Bureau of Land Management, 193 million acres of national forests and grasslands, and approximately 150 million acres of national wildlife refuges[1] may go largely unnoticed, or at least unacknowledged, as equally significant portions of our public land heritage.

This narrow vision is not surprising given that the primary means through which many Americans engage with public lands today is via recreation, something long promoted and prioritized by the National Park Service. Nonetheless, this viewpoint is also highly problematic given the vast array of forces adversely affecting nature and society in twenty-first-century America.

The continued expansion of urban and suburban sprawl, coupled with the early effects of climate change, has stressed or destroyed habitats, threatened native species, degraded water resources, and increased the risk of wildfire on all types of public and adjacent private lands in the wildland-urban interface (or WUI). Economic globalization has led to the decline of both manufacturing and resource extraction industries (e.g., logging, ranching, and mining) that, for better or worse, have impacted public land ecosystems for the past century. This same process has ushered in new amenity- and knowledge-based economies that have drawn urban development and vast numbers of new residents to rural public land communities. Meanwhile, the increased global transfer and consumption of goods has resulted not only in unprecedented levels of global pollution but also growing outbreaks of invasive species and the rapid depletion of resources, all of which have left their mark on the North American landscape. The latter include shortages of water and energy that have pressured managers to develop heretofore untouched public lands. Finally, the concurrent devolution of federal land agencies and the rise of ecosystem-based management have spawned new community-based and collaborative forms of resource management that are challenging conventional understandings of what the "public" in public lands actually stands for.

The cumulative effect of all of this implicates not only national parks, but *all* types of lands and resources in the United States. Conversely, the vast size and environmental diversity of the public domain render it a key resource for mediating the effects of these changes. It therefore behooves us to become better acquainted with the full extent of our public land inheritance: to realize its tremendous value as an environmental bulwark against looming social and ecological crisis, and to learn how to protect our investment so as to pass it on to future generations in better shape than we found it. In short, the time is right to rethink our relationship with public lands and to take a fresh look at our public land heritage.

This book is intended to aid in this task by providing a clear and comprehensive account of America's public lands and resources, including their history, characteristics, current management challenges, and future potential. It is premised on the notion that one cannot fully understand one type of public land without understanding its shared history and relationship to the rest of the system. But it is also premised on the notion that we need to rethink our assumptions and understandings about public lands, including where they came from and what they say about the way we relate to, and value, nature in the United States.

RETHINKING OLD STORIES

A well-worn story of America's public land system in the context of the environmental conservation movement begins this way: Several hundred years ago, European explorers and immigrants came to North America in search of natural resources and to establish colonial empires. After a series of false starts, battles, and negotiations with indigenous peoples and other Europeans, a new nation-state was established. After more battles, conquests, negotiations, and settlement, the land area claimed by that nation stretched across the continent to the Pacific Ocean. As new territory was acquired, it became part of the "public domain." Its purpose? To be sold off as private property for resource development. This usually meant farming and homesteading, but could also include logging, ranching, mining, or some other form of commercial activity.

Over the next one hundred years, the propensity of Euro-Americans to use, develop, and exploit nature fueled enormous economic growth and industrial expansion. But it also rendered significant ecological degradation. Somewhere along the way, a few forward-thinking writers, explorers, poets, and even politicians began to question the dominant course of action. Early American Romantics such as Henry Thoreau and John Muir espoused ideas that challenged the notion of nature as a commodity. Rather, they saw inherent value in nature and argued that it should be preserved and celebrated. Eventually they drummed up enough support to set aside portions of the public domain for federal protection or carefully controlled use via scientific management. The result? The world's first national park. Soon a system of parks, forests, wildlife refuges, rangelands, wilderness areas, and other land types was established to create the nearly 650-million-acre public land system we know today.

The story is powerful and suggests a David-and-Goliath confrontation, with two opposing sides and sets of values facing off over the fate of public lands and resources. The line is drawn between those who want to protect and conserve nature and those who wish to use, develop, and exploit it for profit. Over time, as the battles rage, the pendulum constantly swings between these two poles. Sometimes environmental protections are strengthened, sometimes weakened, but the general framing of the issues remains the same.

This book offers a different narrative. Rather than presenting U.S. public lands as a clean break with the past, I try to show how they actually reflect the *continuity* of certain ways of valuing nature that have been present since the beginning. In this view, the public land system is as much the product of powerful nineteenth-century nature-as-commodity thinking as it is of the early conservation movement. Tracing out the blending of these diametrically opposed ideas does much to explain the tensions, challenges, and contradictions often faced by public land managers today.

Historian and geographer William Cronon made a similar argument in his famous essay, "The Trouble with Wilderness."[2] He wrote that even as environmental groups fought to establish wilderness areas and protect them from development, they continued to adhere to a problematic conceptualization of wild nature shared by their adversaries: wilderness as a place separate from society, characterized as pristine and devoid of human effects. This view was problematic on many levels. It essentially erased from history the impacts of Native Americans upon the landscape. In so doing, it suggested that wilderness was essentially unpeopled, static, and unchanging. This not only set the bar impossibly high for lands to qualify for wilderness designation (allowing opponents to point to almost any human impact as evidence the lands in question were no longer eligible) but also served to diminish the value of nature in which humans reside and use to support their livelihoods. Instead, Cronon urged environmental advocates to take up the call to protect the "wilderness in our backyard": to put more time and effort into finding ways for humans to coexist in nature in more sustainable ways.

While one can take issue with the extent to which wilderness advocates have embraced a "pristine" vision of nature as a precondition for wilderness protection, the notion that long-held assumptions about nature continue to shape and inform our efforts to protect public lands continues to resonate strongly. The legacy of these ideas can be felt throughout the history of the public land system: from the creation of the first national parks (premised in part upon the removal of indigenous peoples[3]), to more recent debates over fire policy on the national forests, the reintroduction of wolves in Yellowstone, or wildlife management on the National Elk Refuge. The ideas of nature-as-commodity, or as something that is "unpeopled, eternal and pristine," continue to render tremendous influence upon the most pressing challenges faced by public land managers today. The manner and extent to which these ideas have come to wield such influence represents a central theme of this book. Understanding them may help to explain one of the greatest conundrums of all: Just how is it that a country more committed to commercial enterprise than any other in the world—and in which private property stands as one of the most cherished national values—chose to set aside almost one-third of its land area as federally managed public lands?

SETTING THE STAGE

The original inspiration for this book derives from *These American Lands* by Dyan Zaslowsky and T. H. Watkins,[4] a text I first discovered while teaching courses on American conservation in the mid-1990s. Back then, it was one of the very few contemporary works that treated America's public land system as a coherent whole, rather than focus on one particular part of the system (e.g., national parks, national forests, etc.), or more commonly, one *specific* park, forest, or refuge.[5]

Since that time, not much has changed. One can readily find books about Yellowstone or the Grand Canyon, but scant attention has been given to public lands as a coherent system. In helping to fill this gap, this book pulls together an extremely wide range of scholarship from multiple disciplines. While I occasionally draw from primary historical sources, including statutes and the published works of historical figures, I also rely heavily upon a wealth of work in natural resource policy, cultural and historical geography, environmental history, environmental sociology, environmental studies, political ecology, and many other disciplines. As before, the lion's share of this research tends to focus more narrowly on one specific type of unit within the public land system. Conversely, I have also benefitted greatly from works that take a more expansive view, treating U.S. public land issues as just one component of much larger narratives of global protected areas[6] or American environmental politics and history.[7]

My own approach to the study of public lands is informed by ideas developed in the field of political ecology; in particular, scholarship that seeks to make explicit the often-taken-for-granted epistemologies or conceptual assumptions that frame our understandings of nature and society, including those deployed in conflicts over the use, control, and meaning of public lands and resources.[8] The end result is a synthesis that brings together diverse insights into a single, coherent narrative, one that offers a critical historical perspective on some of the most pressing resource policy and management issues facing America's public lands today.

Before proceeding, several additional clarifications are in order. The title reads *America's Public Lands* rather than the more accurate phrase, *Federal Lands in the United States*. Formally, the term *public lands* alludes only to those areas managed by the Bureau of Land Management within the Department of the Interior.[9] All other lands in the federal system are officially classified as national forests, national parks, national wildlife refuges, or other categories. Nevertheless, I have adopted the term *public lands* since it remains the most commonly used term for referring to the federal land system. For similar reasons, I use *America* rather than *United States* in the title, understanding that it fails to denote distinctions between North and South, as well as within North America itself. It also bears noting that my focus is on lands and resources managed for conservation or preservation purposes. I do not offer close examination of other types of federal lands (e.g., federal holdings by the Department of Defense, Department of Energy, Bureau of Indian Affairs, and so on), nor of public lands administered by state or local governments.

The book is divided into two sections. Part I examines the common origins of the public domain, including the powerful ideas of nature that continue to shape the public land system to this day. These chapters trace the evolving conceptualization of the public domain from that of a national commons, to a commodity, to the subject of conservation. Specifically, chapter 1 explores the question of acquisition: Where in fact did the public domain come from? Chapter 2 examines our extensive national effort to transfer public lands into private hands. Finally, chapter 3 traces the evolution of the public domain into the public land system we know today.

With this common history in mind, part II explores how these ideas have shaped different types of federal lands within the overall system. With chapters on national parks, national forests, national wildlife refuges, Bureau of Land Management lands, and wilderness areas, respectively, this second part of the book highlights key characteristics and turning points in the development of each land type, including the corresponding federal management agencies. Each chapter concludes with a series of case studies that exemplifies the legacy of certain conceptual assumptions upon current challenges and disputes related to public land policy and management.

Taken as a whole, the chapters are intended to show how the boundaries delineating public lands—both physically and conceptually—have always been more dynamic, porous, and complex than general wisdom suggests. If successful, the book offers not only a fresh historical synthesis but also an opportunity to reconsider our relationship with America's public lands as we move into the next century.

U.S. Public Land System

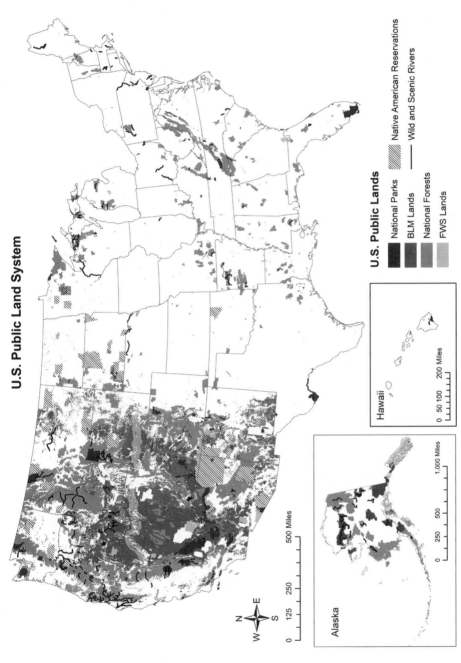

Figure I.1.   The U.S. Public Land System. Map by Jessica Lee.

Part I

ORIGINS OF THE PUBLIC DOMAIN

1

Building the
National Commons

Congress shall have the power to dispose of and make all needful Rules
and Regulations respecting the Territory or other property belonging to
the United States.

—U.S. Constitution, Article 4, Section 3

So just when and where did the public lands originate? If we think of them as a *commons*, an arrangement whereby a group of people share the use and management of land and resources, then the *idea* of the public domain is ancient indeed. Commons-based forms of land tenure stretch back to the earliest forms of human civilization and continue today in many parts of the world. North America is no exception. Prior to European settlement, the commons provided the framework for organizing resources among Native Americans. The use of land for habitation, hunting, gathering, and farming was shared among the members of the local band or tribe. The concept of private property—that is, the exclusive and perpetual use and ownership of land by individuals—only gained ascendancy with the arrival of Europeans.

Though many colonists were drawn to North America by the prospect of acquiring property and thus becoming landowners (something beyond the reach of most living in Europe at the time), they also understood and valued the idea of commonly held lands and resources. The most well-known expression is the "village commons": land used as a shared pasture or for other purposes by local residents.[1] Located in or near the town center, the village commons is a quintessential and persistent part of the New England colonial landscape (think of the Boston Common).

Similarly, in the American Southwest, Hispanic settlers shared water resources as members of *acequias*, or commons-based irrigation organizations, and engaged in *ejido*, usufruct rights to common areas established via Spanish and Mexican land grants.[2]

But these commons, as with many examples across the globe, were decidedly local in character. The shared ownership and use of these lands was limited to members of the local village or community. At the time of the American Revolution, the idea of a *national* commons, or a *national* public domain, had not yet come into being. Creating such an entity would require two things: a modern nation-state and, of course, available land. So in order to answer the question, "Where did the public lands come from?" one must look back to the very founding of the United States itself.

COLONIAL ANTECEDENTS

The story begins with the original thirteen colonies that joined to form the United States, and in particular, the vast swaths of territory to which each laid claim.[3] The original British and Dutch charters for several colonies included clauses granting them all of the land lying east of their northern and southern boundaries, from "sea to sea."[4] Though vaguely defined and often overlapping one another, these charters covered huge expanses of land extending into the unexplored western portion of the continent.

However, by the mid-eighteenth century, the British Crown began to see things differently and moved to place limits on colonial land claims. As part of the treaty agreement ending the French and Indian War, Britain issued the Proclamation of 1763. This law effectively prohibited European settlement between the Appalachians and the Mississippi River—the area that came to be known as the Old Northwest or Ohio River Valley. The rationale given for this decision was to reserve the region in perpetuity as a homeland for Native Americans. It is generally agreed, however, that the actual intent of the Proclamation was to confine European settlement to the East Coast in order to better manage, control, and tax the colonies. For their part, colonial governments tended to ignore the law, offering land to settlers as early as the 1630s as an inducement for immigration and to create a buffer against what they perceived as hostile indigenous peoples.

With the outbreak of the American Revolutionary War in 1776 and the prospect of nationhood looming, the colonies lacking extensive western land claims (e.g., Delaware, Maryland, New Hampshire, New Jersey, Pennsylvania, and Rhode Island) became concerned with the massive claims held by their fellow states. Since land served as the primary form of wealth available to the colonies for paying off war debts and attracting settlers, it was feared that states with huge land claims would quickly gain disproportionate wealth and influence in the new nation. As a consequence, Maryland refused to sign the Articles of Confederation until the western land claims were ceded to the U.S. government as "common stock."[5]

The first draft of the Articles of Confederation was silent on this issue. However, as the war raged on and the tide began to turn against the Americans, several states changed their tune. In 1780, New York became the first state to cede its land claims to the new federal government. The promise by other states to do the same appeased Maryland's concerns and led to the ratification of the Articles of Confederation in 1781. By 1802, all western land claims had been ceded to the nation as a whole. Totaling some 237 million acres, these lands nearly equaled the size of the original thirteen states.

A precedent was thus established for the concept of a national public domain: land owned and managed by, and for, the people of the United States. Federal sovereignty over these lands was further confirmed in a number of ways, including international treaties, early congressional ordinances, and ultimately, the newly minted U.S. Constitution.

Marking the end of the Revolutionary War, the 1783 Treaty of Paris required the British Crown to formally transfer its land claims to the United States. This included the territory stretching from the thirteen colonies west to the Mississippi River, south to the boundary of Spanish Florida (approximately the 31st parallel), and north to the Great Lakes. Britain continued to control Canada, which included all lands north of the 45th parallel to the Pacific Ocean, as well as the area comprising modern-day Oregon, Washington, and Idaho, known as the Oregon Territory.[6]

Two years later, Congress passed the first of two Northwest Land Ordinances. These laws spelled out the purpose of the ceded lands—namely, the creation of new states with the same rights and responsibilities held by other states in the Union. In so doing, the laws also reaffirmed the supremacy of the national government as landowner, requiring that the lands be managed for the "common benefit of the United States."

These powers were again confirmed in the fourth and sixth articles of the Constitution. More specifically, Article 4, Section 3, states that "Congress shall have the power to dispose of and make all needful Rules and Regulations respecting the Territory or other property belonging to the United States." Article 6 establishes the supremacy of federal laws over state laws in cases in which they conflict, as long as the federal laws are in accord with the Constitution. Consequently, as we will see below, when new states were created and admitted to the Union, lands not yet privately owned did not automatically transfer to the states. Rather, the federal government retained exclusive authority over them.

Table 1.1.   U.S. Public Domain: 1781 to 1867

Year(s)	Acquisition	Total Acres (in millions)	% of Total U.S. Land Area	Cost (in million US$)
1780–1802	Original State Cessions	236.8	10.2	6.2
1803	Louisiana Purchase	529.9	22.9	23.2
1782–1817	Red River Basin	29.6	1.3	0
1819	Spanish Cession	46.1	2.0	6.7
1846	Oregon Compromise	183.4	7.9	0
1848	Mexican Cession	388.7	14.6	16.3
1850	Texas Cession	78.9	3.4	15.5
1853	Gadsden Purchase	18.9	0.8	10
1867	Alaska Purchase	378.2	16.3	7.2
Total		1,890.50	79.4	85.1

Source: Based on data from the U.S. Department of the Interior, Bureau of Land Management, *Public Land Statistics 2011, vol. 196* (Washington, DC: GPO, 2012).

Thus, the forging of the national public domain coincided with the creation of the United States itself. The land cessions by the original colonies to the national government set a precedent that would be duplicated many times over in the decades to come. By midcentury, the United States would complete its vision of controlling territory across the entire continent, stretching from the Atlantic Ocean to the Pacific (figure 1.1 and table 1.1). The stories behind this westward conquest are both fascinating and, at times, reflective of one of the saddest chapters in U.S. history. The process was driven as much by political intrigue and economic events in Europe as by the unquenchable thirst for land, wealth, and the fulfillment of Manifest Destiny among U.S. expansionists.

Notably, as with the original colonial land cessions, this westward expansion occurred in multiple phases and within numerous political contexts. On the one hand, it resulted from conflicts, negotiations, and treaties with European empires (e.g., Britain, Spain, France, and Russia) and emergent Euro-American nations (e.g., Mexico and, fleetingly, the Republic of Texas). On the other hand, it involved an additional and separate series of conflicts and negotiations with the indigenous inhabitants of these lands. In what follows, I first recount U.S. interactions with European and Euro-American powers before addressing U.S.-Native American relations.

THE NATIONAL COMMONS EXPANDS

After American independence, the next major acquisition of public land resulted not from U.S. military conquest but the imperial ambitions of

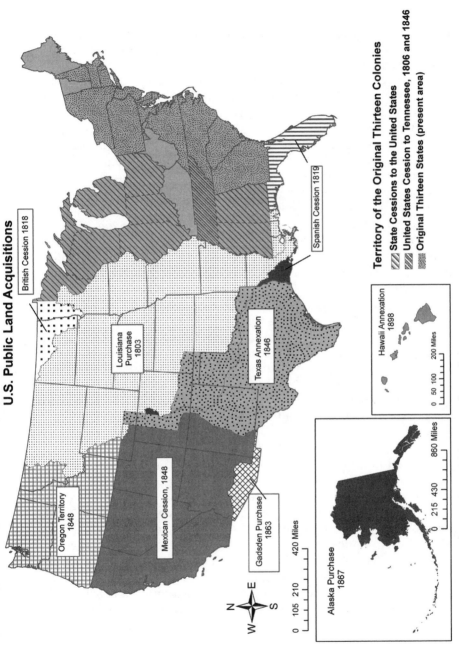

U.S. Public Land Acquisitions

British Cession 1818

Oregon Territory 1848

Louisiana Purchase 1803

Mexican Cession, 1848

Gadsden Purchase 1863

Texas Annexation 1846

Spanish Cession 1819

Territory of the Original Thirteen Colonies

State Cessions to the United States

United States Cession to Tennessee, 1806 and 1846

Original Thirteen States (present area)

Hawaii Annexation 1898

0 50 100 200 Miles

Alaska Purchase 1867

0 215 430 860 Miles

N W E S

0 105 210 420 Miles

Figure 1.1. U.S. Public Land Acquisitions. Map by Jessica Lee.

Napoleon Bonaparte. Circa 1802, France boasted the greatest military power in the world. A declining Spain ceded to France the lands constituting the Louisiana Territory. However, before the French formally took possession, Spanish officials closed the Port of New Orleans to U.S. merchant vessels. Fearing both the permanent loss of a safe port for U.S. goods shipped down the Mississippi as well as Napoleon's plans for a North American empire, President Thomas Jefferson dispatched envoys in 1803 to negotiate for the purchase of New Orleans. The U.S. delegation was authorized to spend $10 million for the city, but by the time they arrived to treat with the French, the situation had changed dramatically.

During the previous six months, Napoleon's North American invasion force in the Caribbean had met stiff resistance in the battle for Santo Domingo and was now severely weakened by disease. With his attention focused on initiating a new campaign against Great Britain, Louisiana was no longer central to Napoleon's plans. Rather than lose the territory to a British counterattack, Bonaparte sought to liquidate his holdings in order to fund conquests elsewhere. Napoleon thus offered to the stunned U.S. envoys not just the city of New Orleans but the entire Louisiana Territory. Eager to seal the deal before Napoleon changed his mind, a treaty was negotiated in less than three weeks, without the authorization of President Jefferson or Congress. Despite heavy congressional opposition over the supposed constitutionality of events, the Senate ratified the treaty in the fall of 1803. According to the agreement, the United States would pay France $15 million for approximately 828,800 square miles of land, an area almost as large as all of western Europe.[7] With the stroke of a pen, the territorial extent of the United States nearly doubled.

However, not everyone was pleased with the arrangement. The vaguely defined western boundaries of the vast Louisiana Territory quickly led to tensions with Spain. The Spanish interpreted Louisiana's boundaries as limited to the cities of New Orleans, St. Louis, and a narrow strip of land along the western bank of the Mississippi River. In contrast, the United States viewed Louisiana as encompassing all lands west of the Mississippi, stretching south to the Rio Grande and west to the Rocky Mountains, including the northern headwaters of the Missouri River.

Unfortunately for Spain, things were not going well in the early part of the nineteenth century. Facing a trifecta of problems—political unrest at home, the threat of French invasion, and the specter of rebellion in its New World colonies—Spain was in no position to risk military confrontation with the United States. As a consequence, Spain promptly signed the Adams-Onís Treaty of 1819. The agreement required Spain to sell Florida and the lands on the eastern shore of the Mississippi Delta to the United States for approximately $5 million. The Treaty also set the western boundaries of the Louisiana Territory at the Sabine, Red, and Arkansas rivers, leaving Spain with modern-day Texas and all lands west of the Rockies below the 42nd parallel (the northern boundary of modern-day California). Finally, as part of the agreement, both Spain and Russia relinquished their claims to the Oregon Territory.

But not all territorial gains were so peacefully negotiated. In 1835, a group of American immigrants in Texas, who had been welcomed by the Mexican government to help settle the northern portion of the territory, rebelled against the authoritarian regime of President Santa Anna. In 1836, after months of battle, a coalition of Anglo and Spanish-Mexican residents successfully seceded from Mexico, establishing the independent Republic of Texas. Despite early efforts to join the United States, the annexation of Texas did not occur until 1845. As part of the deal for statehood, Texas agreed to cede over seventy-eight million acres to the national government. Today these lands constitute portions of New Mexico, Colorado, Oklahoma, and Kansas.

The annexation of Texas also held significant implications for the next two major episodes of U.S. westward expansion: the Oregon Compromise and the U.S.-Mexican War. The year prior, in 1844, James Polk won the U.S. presidential election running on an expansionist platform. Firmly committed to the idea of Manifest Destiny, Polk believed the United States had a moral responsibility to colonize the entire North American continent. Earlier treaties signed with Britain, in 1818 and 1827, respectively, had postponed the question of control over the Oregon Territory by establishing a policy of joint occupation. Eager to rectify this unsatisfactory and temporary resolution, U.S. and British government officials agreed to discuss the matter. The territory in question was vast, stretching from the Missouri River to the Pacific Coast, and from the northern boundary of California at the 42nd parallel to the southern boundary of Russian-controlled Alaska at 54°40'. With U.S.-Mexican relations rapidly deteriorating over the annexation of Texas and the British preoccupied with imperial pursuits elsewhere, the delegates agreed the best solution was to partition the territory at the 49th parallel. Through the Oregon Compromise of 1846, the United States thus acquired over 183 million acres of land, constituting modern-day Oregon, Washington, and Idaho.

That same year, the United States entered into a two-year war with Mexico. On the Mexican side, aggression was triggered largely by U.S. support for the secession and later annexation of Texas, two actions that Mexico had never recognized. In the United States, the most vociferous proponents were expansionists who viewed military action as the best chance of acquiring long-coveted Mexican lands in the West. The war ended with the defeat of Santa Anna's forces and the occupation of Mexico City by the U.S. military. In the 1848 Treaty of Guadalupe Hidalgo, over 388 million acres were transferred to the United States, including modern-day California, Nevada, Utah, and portions of Arizona, Colorado, and Wyoming. In return the United States paid Mexico $18 million and agreed to respect existing property and land grant claims by Mexican citizens—though in practice these promises were frequently ignored. Five years later, the United States once again approached the Mexican government seeking additional land to enable the construction of a southern transcontinental railroad line (what would become the Southern Pacific). Under the terms of the resultant Gadsden Purchase of 1853, the United States paid $10 million to Mexico in exchange for approximately nineteen million acres in southern Arizona.

The final significant expansion of the public domain in North America took place in the northernmost portion of the continent. In 1858, fearing that Great Britain was about to take its Alaskan Territory by force, the Russian Empire put the land up for sale as a preemptive measure. With its initial offer to Britain rebuffed, in 1859, Russia extended an offer to the United States. But the outbreak of the U.S. Civil War delayed any decision on the matter. Finally, in 1867, with the war over, the Russians once again offered the territory to the United States. This time, through negotiations led by Secretary of State William Seward, the United States agreed to buy 586,412 square miles of land for $7.2 million, or about 1.9 cents per acre. Criticized by some at the time, including newspaperman Horace Greeley, as "Seward's Folly" due to the perceived worthlessness of the land, the purchase turned out to be one of the greatest real estate bargains in history.

While the acquisition of Alaska marked the end of U.S. expansion on the North American continent, the process continued overseas.[8] Proving the adage that truth is stranger than fiction, Congress passed the Guano Islands Act back in 1855. Under this law, the United States unilaterally granted itself the power to control islands in the Pacific Ocean containing guano (also known as bat droppings) as a source of saltpeter for making gunpowder. Justifying this action as a matter of national security, the United States consequently claimed over one hundred small islands throughout the Pacific. Today, though some remain disputed, the United States continues to claim eleven islands under this 1855 provision, including the Northern Mariana Islands, which formed a Commonwealth Association with the United States in 1975.[9]

Other overseas land claims followed a more conventional imperial logic. In 1898 Hawaii was annexed to the United States as a territory following the 1894 overthrow of the indigenous monarchy through a coup facilitated by American attorney Sanford Dole, a cousin of pineapple magnate James Dole. That year also marked the end of the brief Spanish-American War, which had been fueled by U.S. support for Cuba's rebellion against Spanish rule. The resulting treaty gave the United States control of Cuba, the Philippines, Puerto Rico, and Guam. While Cuba (minus Guantanamo Bay) and the Philippines later gained independence, Puerto Rico and Guam remain territories of the United States to this day.

The following year, the United States acquired a portion of the Samoan Islands. To address competing claims by the United States, Germany, and the British Empire for control of the islands, the Tripartite Convention of 1899 was held to settle the matter. After Britain agreed to relinquish its claims, the archipelago was partitioned into a German colony and the U.S. Territory of American Samoa.

Eighteen years later, the United States made its final overseas acquisition. Claiming that the Virgin Islands were essential for defending the Caribbean against German submarine attacks during World War I, the United States convinced Denmark to sell them in 1917. Now known as the U.S. Virgin Islands, they include St. Croix, St. Thomas, St. John, and several smaller islands in the region.

THE FEDERAL INDIAN RESERVED LANDS

In a process parallel to the negotiations and wars with European and Euro-American powers, the acquisition of the public domain required a separate set of interactions with the indigenous inhabitants of North America. The often contradictory, poorly enforced, and ill-conceived attempts to broker and implement land purchases, exchanges, and treaty agreements between the United States and Native American tribes mark one of the saddest chapters in U.S. history. They constitute a legacy that continues into the present and, like the pubic domain, has roots that can be traced to the very creation of the United States itself.

Recall the 1783 Treaty of Paris? The one in which the British surrendered their colonial land claims to the United States at the end of the Revolutionary War? While explicit about U.S.-British territorial claims, the agreement required the new American government to address these issues separately with indigenous peoples. Treaties signed in the 1780s that divided lands between Native Americans and white settlers repeatedly failed, foreshadowing the next two centuries of U.S.-Tribal relations. The treaties suffered from a variety of shortcomings: misunderstandings regarding the terms and conditions of the documents, the lack of full participation by all tribes in the area, false presumptions that the indigenous leaders who signed the agreements in fact spoke for others within the tribe or nation, the absence of U.S. government enforcement regarding settlement restrictions, and the lack of recognition of indigenous land rights by white settlers and land speculators who continued to flood into the region. The issue was ultimately decided through a decade of military violence, culminating in the pivotal Battle of Fallen Timbers in 1794. The following year, twelve tribes surrendered their territorial claims to the United States in the Treaty of Greenville.[10]

The devolution of relations with Native Americans in the Northwest Territory represents a pattern repeated again and again in U.S. history: precursory treaty negotiations and agreements, followed by numerous renegotiations, the erosion of relations into armed conflict, and ultimately, the forced removal of indigenous people to reservations. And rarely did matters end there. Over time, interest in these Federal Indian Reserved Lands, as they are called, was renewed, often for purposes of mining development, speculation, or settlement. These actions triggered new rounds of conflict between white settlers and Native Americans. Ultimately, such events led to further attempts to remove tribal members altogether from their homelands or to reduce the size of reserved areas to ever smaller parcels of marginalized lands.

The Oklahoma Territory is a case in point. Originally created as a permanent Indian Territory through the Indian Removal Act of 1830, tribes from western Georgia were given the choice of moving westward and retaining their sovereignty or remaining in Georgia but losing tribal recognition. The result was the famous Trail of Tears, in which thousands of Native Americans died on the march to Oklahoma. Then, in 1887, the Dawes Act broke up the communally owned land system, forcing tribal members to become individual property owners and opening additional land to white settlement.[11]

Put simply, the lands to which indigenous peoples were removed—the Federal Indian Reserved Lands—were and remain a contradiction in terms. On one hand, the tribes are sovereign nations with the power to govern themselves and their territory. When the governmental authority of tribes was first challenged in the 1830s, U.S. Supreme Court chief justice John Marshall ruled that the tribes *possess "nationhood status and retain inherent powers of self-government."*[12] On the other hand, the tribes have never been recognized as such by the U.S. government in the same way as other sovereign nations. Nor are they allowed to make laws that supersede federal laws, establish a military, or issue their own currency. To further complicate matters, the U.S. government serves as a "caretaker," holding the reserved lands in trust for the tribes via the Bureau of Indian Affairs. And, as of 1924, all tribal members born in the United States are also U.S. citizens.

Currently, there are some 564 recognized American Indian tribes and Alaska Natives in the United States, with a combined population of approximately 1.9 million people. There are 326 federal Indian reservations, covering approximately 56.2 million acres.[13] They range in size from the sixteen-million-acre Navajo Reservation in Arizona, New Mexico, and Utah, to the tiny 1.3-acre cemetery of the Pit River tribe in California.

While it can be argued that the Federal Indian Reserved Lands are also public lands, their raison d'etre and complex, quasi-sovereign status sharply distinguish them from other lands within the public domain.[14] A full, detailed account of these lands is well beyond the scope of this book. Nonetheless, I briefly touch upon U.S.-Tribal relations here and elsewhere insofar as they directly relate to the acquisition and later disposition of the public domain. I also attempt to show how these relations have played a significant if largely unacknowledged role in shaping the way we have come to perceive, value, and use our public lands, as explained in later chapters.

SUMMING UP

The history of public land acquisition in the United States contains the seeds of two major conceptual assumptions relevant to the challenges facing public land management today. The first assumption is the authority of the national government over these lands. Native American claims notwithstanding, this authority stretches back to the very foundation of the United States itself, a fact that is salient in light of more recent claims that the public domain represents lands that were somehow wrongfully taken away from states, counties, or municipalities. Periodically over the years, advocates for privatizing the public domain have made such claims. The Sagebrush Rebellions of the 1940s and 1970s espoused these views, as have members of the so-called County Supremacy and Wise Use Movements in more recent years.[15] While it is true that some lands currently managed by federal agencies were purchased from or donated by private landowners or state governments—sometimes under declarations of eminent domain—the idea that the federal government lacks authority over the public domain simply does not hold water.[16]

The supremacy of federal authority is also relevant for other reasons; namely, the unique conceptual challenges it poses for the adaptation and implementation of collaborative or community-based forms of resource management. In recent years, finding ways to accommodate the voices, knowledge, and views of various local residents and governmental authorities without compromising the decision-making sovereignty of the federal government over the public domain has been nothing short of daunting.[17]

A second important conceptual assumption is the often-unrecognized legacy of the forced removal of indigenous peoples from the public domain. For decades, the dominant narratives of the U.S. conservation movement argued for the creation of national parks, forests, and wilderness areas on the premise that these portions of the public domain represented something akin to pristine nature. But the only way these places could possibly be "pristine" is by simultaneously viewing them as devoid of human influence, wiping away the history of human occupation prior to European settlement. This leaves us with a static vision of the seventeenth-century North American landscape as the quintessential example of untouched wilderness, a vision that ignores the impacts of Native Americans in shaping those landscapes, from the composition of forests to the expanse of prairies and the abundance of wild game. As William Cronon and others have shown, this way of seeing the land is not only historically inaccurate but also rendered powerful impacts upon the way we would later come to understand wilderness protection: remove humans and the land would remain forever pristine.[18] Such views encouraged a related and equally problematic notion of nature as static and unchanging. With the advent of ecosystem management and the concept of sustainable development in the 1980s and 1990s, the legacy of these ideas—of nature as "unpeopled, eternal and pristine"—complicated attempts to envision just how human activities and livelihoods might be meaningfully integrated into land and resource management in more sustainable ways.

As we will see, the legacy of these conceptual assumptions continues to inform public land management debates today. But first it is important to explore how these ideas influenced the next stage in the history of the U.S. public domain: selling it off!

2

Disposing of the Public Domain

From Commons to Commodity

In Europe the lands are either cultivated or locked up against the cultivator. . . . But we have an immensity of land courting the industry of the husbandman. . . . Those who labor in the earth are the chosen people of God.

—Thomas Jefferson[1]

The untransacted destiny of the American people is to subdue the continent . . . to rush over this vast field to the Pacific Ocean . . . to confirm the destiny of the human race.

—William Gilpin[2]

Once acquired, what was the purpose of the lands within the public domain? Both the Constitution and the Land Ordinances of 1785 and 1787 are clear that the lands were intended for the creation of new states. Moreover, they allotted sole authority over this process and the public domain itself to the national government. Article 4 of the Constitution specifically directs the government to "dispose of and make all needful rules and regulations" deemed necessary for this task. Key to this mandate is the word *dispose*. To Americans in the late eighteenth and early nineteenth centuries, disposing of the public domain did not mean throwing it away, but rather selling it as private property. Thus, concurrent with the creation of the United States, the federal government assumed the power to sell public lands, transferring the newly acquired national commons into private hands. As historian Richard White observes, the role of the federal

23

government vis-à-vis the public domain during this period is best characterized as that of a real estate agent rather than a landlord.[3]

But did this mean that as new states were carved out of the public domain, all unclaimed lands lying within their borders would be transferred to state governments as well? The answer is a clear and resounding "no," according to early congressional land laws. The Northwest Ordinance of 1787 explicitly prohibited states from getting involved in the business of selling the public domain. Nor were states allowed to tax existing federal lands within their borders.

So why—once states were created—was it so important for the national government to retain control over public lands, including their transfer into private ownership? One word: money. Like the original colonies, land represented the most valuable commercial asset owned by the United States in the wake of the Revolutionary War. For the cash poor national government, public lands were the primary source of revenue for paying off debts and stimulating economic development.

This was done in a number of ways. For example, to pay war debts, the government issued military land bounties. Plots of private property ranging from fifty to eight hundred acres each were given as rewards for military service, a practice that continued after the War of 1812 and the U.S.-Mexican War in 1848. It is estimated that some sixty-one million total acres of land were dispersed to private individuals in this way.[4] Land bounties and grants were also an effective means of inducing westward settlement (especially in the expansionist 1840s), generating funding for schools, revenues for state governments, and investment capital for large-scale commercial improvements such as roads, canals, and railroads. And of course, selling land to individual settlers created a revenue flow that fed directly into the coffers of the national government.

But just *how* was the land to be sold off? The process of disposal became a hotly contested issue among the nation's leaders, intertwined with larger debates about the future direction of the U.S. economy, the political system, and the role of the federal government vis-à-vis the states.

PRIVATIZING THE COMMONS: TWO VISIONS

In the 1790s, then as now, the political landscape in the United States was divided into two competing national parties. On one side stood the Democratic-Republicans, led by Thomas Jefferson and James Madison. They favored a strict interpretation of the Constitution, strong state rights, and a relatively small central government. Jefferson was particularly concerned that the United States should break away from European traditions in which landownership was the prerogative of the wealthy elite while the majority lived as poor tenants. In contrast, Jefferson envisioned a nation of "yeoman farmers": small, independent landowners representing the backbone not only of the American economy but its political system as well. This was Jefferson's idea of Agrarian Democracy.[5] According to this logic, a properly functioning democratic government required two essential ingredients: the diffusion of political power away from eastern urban elites, and a sufficiently independent electorate able to make political decisions on merit rather than be swayed by powerful economic benefactors. The vast expanse of the public domain provided the necessary land resources for such a vision. Jefferson thus proposed that the public lands be sold in relatively small plots, at prices affordable for individual settlers.

On the other side of the political divide stood the Federalist Party, founded by Alexander Hamilton and James Monroe. Federalists advocated for weaker state powers and a strong central government. They also believed that first and foremost, the public domain should serve as a resource for generating revenue. Accordingly, public lands should be sold in large chunks to groups of investors or corporate enterprises who could most effectively develop the resources therein for large-scale agricultural production, the extraction of raw materials for manufacturing, or for the purposes of land speculation. Such actions, it was argued, would support the development of a diversified national economy and allow the United States to more quickly "catch up" to its European rivals.

The earliest congressional statement on this issue was the Land Ordinance of 1785. After purchasing Native American claims to the land, the law required federal surveyors to mark it off into giant thirty-six-square-mile plots called townships. Each plot was further divided into one-square-mile sections (totaling 640 acres each). The land was then sold at auction to the highest bidder at a minimum price of $1 per acre (raised to $2 per acre in 1786, then reduced to $1.25 per acre in 1820). Nonetheless, within every township five lots were reserved: sections number 8, 11, 26, and 29 were held by the national government to generate revenue for state or territorial governments, and section number 16 was set aside to provide funding for local schools. Later, an additional section was assigned to further support schools. The resulting checkerboard pattern of the township system can still be seen today, imprinted on the agricultural landscape of the Midwest (figure 2.1).

Congress adopted this rectilinear township system from the New England colonies. Viewed as a rational and orderly way to dispense the public domain, it contrasted sharply with the "metes and bounds" system used in the South, in which settlers defined their own boundaries using natural landmarks. The resulting irregular patchwork of parcels made administration difficult and led to numerous boundary disputes, especially when markers such as riverbanks or trees shifted or were cut down.[6]

Figure 2.1. An aerial view of Midwestern farmscapes illustrates the legacy of the rectilinear survey system used to dispose of the public domain. Photo courtesy of Airphotona.

Yet despite its geometric elegance, the township system also contained significant flaws. To begin with, settlement tended to outpace the survey and auction process. As a result, by the time land was surveyed and finally put on the auction block, it was not uncommon to find it already occupied by squatters. Second, the township system ignored geography. While neat and orderly for record-keeping purposes and boundary delineation, the imposition of an abstract rectilinear framework over the North American landscape contrasted sharply with ecological realities on the ground. As we will see, the jurisdictional and conceptual legacy of this development would haunt environmental resource conservation efforts well into the future.

But to return to our original question: Which of the two competing political visions for land distribution eventually won out? Was it the Jeffersonian approach favoring small landholders, or the Federalist philosophy giving preference to wealthy investors? It turns out that the answer changes depending on the time period in question. Early laws clearly favored the Federalist position. Laws passed in 1785 required buyers to either purchase entire townships (thirty-six square miles or 19,840 acres once the five sections were withheld) or individual 640-acre sections (but only in alternating townships). Very few settlers could afford to purchase a 640-acre section of land, much less an entire township. Moreover, the land offices set up to handle transactions were located in eastern cities, far from the actual sites of settlement. The next relevant legislative action on the matter—the Land Law of 1796—offered more of the same. One could now purchase 5,120-acre "quarter townships" and use credit for one year with a 5 percent down payment. However, the minimum purchase price was raised to $2 per acre. In 1800, the Harrison Frontier Land Act extended credit for land purchases to four years, but it changed little else.

Not surprisingly, none of the early public land laws triggered a rush to settlement in the public domain. However, things changed after Jefferson won the presidency in 1801. With the acquisition of the Louisiana Territory in 1803, a series of new laws and regulations were issued that favored Jefferson's vision. Land prices became more affordable to individual settlers and wealthy investors alike.

A new law passed in 1804 extended credit to six years, allowed a lower minimum purchase of 160-acre quarter-sections, and opened new land offices near the areas being auctioned off.[7] These changes sparked a land rush that intensified again after the War of 1812. Overwhelmed by the mammoth task of surveying and administering the sale of public lands, in 1812 the government established the General Lands Office within the Department of Treasury.[8] Yet by 1819, the boom had collapsed as the overproduction of agricultural goods drove down prices. This led to a series of laws that continually tinkered with public land policies, alternately raising and lowering minimum purchase prices and credit limits. According to some estimates, between 1789 and 1834, Congress passed 375 land laws dealing with different ways through which the public domain could be transferred into private hands.[9] At the same time, government enforcement of these laws remained weak or nonexistent, opening the door to rampant abuses by speculators and emergent resource extraction industries.

Significant among the new laws were those dealing with the huge number of illegal squatters on the public domain. In the 1830s, several temporary Preemption Acts were passed, giving squatters first right to buy the land upon which they had illegally settled. As long as they made improvements to the land, they could buy up to 160 acres before the auction for the minimum asking price (i.e., preempt it). Passage of a permanent law in 1841 effectively ended the auction process altogether, opening the entire domain to settlement through the staking of claims.

Other efforts to encourage the sale of public lands included the Graduation Act. Following the logic that unclaimed land was less valuable than other parcels in the public domain, in 1854, Congress passed this law that reduced the minimum price of land to $1 per acre if unclaimed after ten years. After thirty years, the price was further reduced to 12.5 cents per acre. The net effect was an upsurge in Midwestern settlement in the mid-nineteenth century.

THE HOMESTEAD ACTS, LAND GRANTS, AND RAILROADS

The Civil War years witnessed yet another round of laws designed to dramatically impact the distribution of public lands. With Congress now completely controlled by the Union, northern politicians actively sought to promote the Jeffersonian ideals of Agrarian Democracy in the West as a guard against the expansion of large-scale plantation agriculture and, by extension, slavery. To this end, in 1862, Congress passed the Homestead Act and the Morrill Act, and initiated the first of many transcontinental railroad grants.

Arguably, the Homestead Act marked the final victory of Jefferson's vision of a nation of small free-holder farmers. Under the terms of the act, 160 acres of land was basically given to settlers for free, as long as they fulfilled the requirements of residing on and improving the land for five years and paid a $26 fee (amounting to 16.25 cents per acre). Alternatively, land could be purchased for $1.25 per acre after only six months of inhabitation. At the time, relatively few took up the offer. Between 1862 and 1900 approximately one to four million acres per year were homesteaded. The greater impact took place between 1908 and 1922, when seven to ten million acres per year were transferred into private hands due to new homestead laws that expanded claim sizes from 160 up to 640 acres to facilitate dry-land farming and stock raising.[10] Though roughly one-third of homesteaders failed to ever complete their claims, the Homestead Acts still resulted in privatizing some 287.5 million acres.[11]

Also passed in 1862, the Morrill Act helped to ensure the success of the Homestead Acts. It did so by providing land to the states to finance higher education, leading to the establishment of the land grant university system. Notably, the universities focused instruction on applied fields of study, including the mechanical arts, home economics, and agriculture. The laws' sponsors hoped to make education available to a wider array of social classes, including those pursuing blue-collar livelihoods.[12]

During this period a third type of land law emerged, derived from a completely different set of American ideals. While the Homestead and Morrill acts emphasized Jefferson's vision of westward settlement by individual farmers, the Federalist vision of using the public domain to support large-scale capital investments and a diversified economy found its own expression in federal grants to the railroads.

Significantly, the railroad industry benefitted from both visions of America's future. For homesteaders, railroads offered access to the public domain for settlement and transportation for their agricultural goods to eastern markets. For large capital investors, railroads enabled the extraction and transport of western raw materials to eastern manufacturing centers, all the while providing ample opportunity for land speculation.

Meanwhile, Congress also looked for ways to offer financial support to the industry. For northern politicians, the railroads represented a means of tying the western territories more closely to the Union. These were heady times indeed for those in the railroad business.

Federal transcontinental railroad grants followed the formula established by earlier grants to the states for building roads and canals. In the first years of the nineteenth century, the national government gave 5 percent of revenues from the sale of the public domain to the states, with 2 percent of those funds earmarked for transportation projects. However, as early as 1808, the national government gave portions of the public domain as rights-of-way to the states for the construction of roads. Over the next decade, it did the same for canals. In 1827, the United States initiated an additional strategy: granting land to construction companies for investment purposes. By reselling the land, the company could use the profits to finance construction of the canals. The national government then recouped its losses by selling the adjacent land parcels at twice their normal price, under the assumption that canal improvements would raise land values. Generally speaking, the canal companies were given 640-acre sections on alternating sides of the route within a distance of five miles for the entire length of the canal. In instances where the property was already occupied, lands could be granted up to fifteen miles away from the canal route.

Federal railroad grants included the rights-of-way and ten alternating 640-acre sections on either side of the tracks per mile. In addition, Congress allowed railroad companies to take all needed timber and stone from the public domain free of charge and provided thirty-year government loans ranging from $16,000 to $48,000 per mile depending on the terrain.[13] The deal was even better for transcontinental railroads. Instead of ten sections, they received twenty sections per mile and could select "lieu lands" anywhere they wished on the public domain. The largest grant in history went to the Northern Pacific, which received twenty sections per mile through states and forty sections per mile through territories. By the end, this amounted to nearly forty million acres transferred to a single private company, an area the approximate size of New England.[14] In total, between 1862 and 1872, the federal government gave over ninety-four million acres of public lands directly to the railroad industry, and another thirty-seven million acres via the states (see table 2.1).

Notably, these land grants were not to be held indefinitely by the railroad companies. Congress intended the companies to turn around and sell land grant properties relatively quickly. Depending on the specific grant, firms were given between three and five years after the completion of the line to sell the land. However, in practice, companies often held the land much longer, often delaying decisions on the right of way for years. By withholding millions of acres from homesteading, land prices rose dramatically, limiting opportunities for settlement and frustrating western farmers, land developers, and politicians. In 1871, these tensions finally led Congress to terminate the federal railroad grant program.

One major problem with both the Homestead Acts and the railroad grant programs was the fact that they both treated the public domain as an abstracted, undifferentiated landscape. While railroad companies sought out the best routes and selected the most commercially valuable parcels, and homesteaders and speculators attempted to claim lands with superior soil conditions and access to water resources, the national government, for its part, continued to value all land as though it was equal. Following the logic of the rectilinear survey system, the grant programs ignored geography. Except for lands adjacent to transportation improvements, all portions of the public domain held the same monetary value in the eyes of the government, whether located in a fertile river valley, an arid desert, or on a mountain peak.

This same logic carried over to questions of land *use* as well. Federal laws operated under the assumption that the highest and best use of the public domain was either for mining or agricultural development as practiced in the eastern United States. Accordingly, the 160-acre allotments in the Homestead Act were premised on the average farm size east of the Mississippi River, where soil and climatic conditions allowed them to prosper. But whereas 160 acres constituted a viable farming operation in Pennsylvania or Ohio, it did not suffice in the arid conditions of western Kansas or Arizona. As the number of failed homesteads began to soar, it became clear that a new perspective was desperately needed.

LOGGING, RANCHING, AND MINING

By the mid to late 1800s, with much of the best agricultural land already taken, growing numbers of homesteaders found themselves forced to stake claims on increasingly marginal plots. Not surprisingly, these settlers frequently failed to "complete" their claims according to the requirements of the Homestead Acts. At the same time, the public domain became subject to new forms of resource development largely unaccounted for in existing land laws. The most prominent among these included mining, logging, and ranching. In response, Congress passed a series of new land laws designed to acknowledge—if not adequately regulate—these new uses and values.

Some of the earliest examples of these new statutes related to wetlands and floodplains—places considered worthless in terms of agricultural production. Through the Swampland Acts of 1849, 1850, and 1860, Congress transferred these parcels to the states, following the model set by the state canal and railroad grants. States were required to sell the wetlands to investors in order to fund their draining, the building of levees, and other improvements. This would increase the amount of arable land available for settlement and reduce the threat of floods. In keeping with other public land laws of the period, the Swampland Acts completely ignored the ecological role of wetlands and floodplains in preventing or mediating the effects of flood events. Moreover, the laws were open to rampant abuse by state officials who used rather loose definitions of "swamp and overflowed lands unfit for cultivation" in order to gain control over more attractive federal lands. Ultimately, the Swampland Acts enabled the transfer of over sixty-four million acres of public land to the states.

Mineral resources represented another value generally unaccounted for on public lands. Gold and silver rushes in California in 1848, Nevada in 1858, and Colorado in 1859 created new demands for public lands and resources that had nothing to do with settlement or agricultural production. In the immediate aftermath of these rushes, miners developed an extralegal system for land tenure based on the concept of prior appropriation: those with first claims had the strongest rights. In 1866, Congress passed the first mining law, reaffirming much of this system. It stipulated that mineral lands in the public domain were to be "free and open to exploration and occupation." Lode mining claims could be purchased for $5 per acre as long as the land was occupied and the claimant had invested $1,000 worth of labor and "improvements." The 1870 Placer Act established similar rules for surface mining claims, allowing miners to preempt 160 acres for $2.50 per acre, with no limits on the number of separate 160-acre claims made per individual.[15]

The 1872 General Mining Law combined these two laws and is still in effect today. As the final word on the matter, it restricted land claims to "valuable" rather than "all" mineral lands, reduced the amount of investment required to $100 worth of improvements per year, and extended its application to "all lands owned by the United States government." Significantly, the law did not require royalties to be paid to the U.S. government on the minerals extracted. Though the law was intended for individual miners, the main beneficiaries tended to be large mining companies, both foreign and domestic. Once the more easily accessible surface minerals were depleted, only large-scale enterprises could afford the machinery and additional infrastructure needed for extracting and processing mineral deposits located at greater depths.

Perhaps the most far-reaching element of the General Mining Law was the "patent" clause. Once miners invested $500 of improvements, the law allowed mining claims to be "patented" (e.g., purchased) as private property and put to any use deemed appropriate by the owner, not just

mineral development. This created a powerful incentive for land speculators to acquire portions of the public domain under the guise of mining,
but with the intention of using it for other purposes once patented. As a
result, many types of federal lands today are littered with thousands of
small pockets of private property known as inholdings. Such places frag-
ment habitats and greatly complicate efforts by federal agencies to manage public lands in a coherent fashion.

In addition to the General Mining Law, another set of statutes attempted to deal with the problem of establishing homesteads and farms
in arid conditions out West. Following the belief that "rain follows the
plow," Congress passed the Timber Culture Act of 1873. This law allowed
settlers with 160-acre claims to receive an additional 160 acres as long as
they promised to plant trees on forty of them (later reduced to ten acres).
The idea was that the new trees would change the climate and induce
precipitation. While we know today that large, mature forests can create
microclimates by altering temperatures a few degrees, merely adding
forty acres of young trees per 160 acres of arid land would not render any
such effect upon precipitation patterns. The notion is likely based on early
observations that forested areas receive greater rainfall. But such logic reflects a misunderstanding of cause and effect. Without water, most young
trees were doomed from the start. Consequently, most claimants under
this Act either failed utterly or never entertained any real expectation of
complying with the law in the first place. Rather, due to lax enforcement,
settlers and speculators could skirt around the requirements but still
double the size of their original homestead claims.

Three years later, Congress tried to tackle the problem of aridity once
again, this time providing incentives for irrigation. The 1876 Desert Lands
Act allowed the purchase of 640-acre claims for $1.25 per acre as long as
the owner promised to irrigate the land within three years. Again, while
the law represents a belated acknowledgment of the arid conditions in the
West, it reflects a misunderstanding of ecology and the near-impossible
task of individuals to successfully irrigate desert lands on such a scale.
Like the Timber Culture Act, this Act reflected the continued belief that
agriculture was not only possible but remained the best use of western
public lands. And also like the Timber Culture Act, the Desert Lands Act
was a monumental failure in achieving its stated objective.

But this did not mean the law was without significant and lasting regional impacts. In fact, the Desert Lands Act did much to facilitate the
creation of vast cattle ranching empires in the desert Southwest. While
not suitable for farming, the arid and semiarid conditions of the western
public rangelands were ideal for livestock grazing. Ranching, however,
required thousands of acres of land, well beyond the hundreds made
available through the Homestead, Timber Culture, and Desert Land

Acts. To multiply the effects of these laws, ranch owners often relied on proxies: usually ranch employees and family members, who would file claims on land near to or containing water resources. The rancher paid for any required tree plantings or irrigation "improvements," which might amount to nothing more than a single shallow ditch, then provided the proxy with cash to clear the title. Finally, the proxy would "sell" the land back to the rancher. By owning all parcels of land containing water resources in an area, a ranch owner rendered the remaining vast tracts of arid land virtually worthless for homesteading. Thus by default, thousands of additional acres of public lands fell under the control of large ranching operations (see chapter 7).

A final set of laws addressed the question of timber resources in mountainous regions of the public domain. Like swamplands, deserts, and lands rich in mineral resources, as long as the land had been surveyed for homesteading, the Homestead Act and other early land laws treated forested lands as though they were no different from any other part of the public domain. This created incentives for logging companies and land speculators to file claims in forested areas, remove the trees, then either abandon the claims (simply by failing to pay taxes) or resell the resultant denuded landscapes to potential farmers.

In an ill-fated attempt to rein in this practice, Congress passed the Timber and Stone Act of 1878. Under this law, the government could sell public lands that were deemed unfit for agricultural or valuable mineral production but that still held some value for the timber or building stone contained therein. This allowed homesteaders with 160-acre claims to purchase an additional 160 acres of such lands for $2.50 per acre, a price that was less than one-tenth of the estimated timber value at the time in places such as the coastal redwood forests.[16] The catch was that the trees could only be harvested for personal use. However, again, due to lax enforcement, large timber companies used proxies to make multiple claims under the Act. In this way, they came to control vast expanses of forested land. The companies then removed the trees illegally, leaving degraded landscapes ripe for wildfire and flooding in their wake. In a related law, the 1878 Free Timber Act, Congress allowed for a similar personal use of timber (e.g., for agricultural, mining, or other domestic purposes) on lands designated exclusively for mineral development.

FEDERAL INDIAN RESERVED LANDS REVISITED

The story of public land disposal for settlement did not end with the Homestead Acts, railroad grants, and other nineteenth-century land laws. Long after many of these laws had been repealed or amended, the hunger

Chapter 2

for land remained insatiable. Toward the end of the 1800s, with much of the perceived "best land" for settlement already taken, Congress turned to the Federal Indian Reservation system in search of more.

In 1887, sponsors of the Dawes Act successfully argued in Congress that the reservations, which represented land owned in common for all members of a specific tribe or nation, were in fact too large and impeded cultural assimilation. However, if Native Americans were to become private property owners instead—with 160-acre plots per head of household and smaller allotments given to each tribal member—they could learn to be self-reliant farmers and play a greater role in mainstream society. And more importantly, the logic went, the remaining reservation land could be opened to additional white settlement. According to the law, individual property allotments would be held in trust for tribal members by the U.S. government for twenty-five years, but if the tribal members so chose, these could be sold off as well.

Through this set of rationales, the Dawes Act gave the U.S. president power to break up commonly held reservations. Initially, certain eastern tribes were exempt, but later legislation eventually erased these distinctions and encompassed all federal Indian reservations. Given the unsuitability of most individual lots for farming and the wish of many tribal members to continue their traditional ways of life (to the extent this was possible on the reserved lands), the Dawes Act was yet another disaster for Native Americans. Although the law stated that selling the "excess" lands was voluntary, the influence of monetary incentives for tribal leaders, miscommunication, and a healthy dose of corruption among white government officials and land speculators led a number of tribes to forfeit their common lands. In Oklahoma, between 1891 and 1901, the law resulted in no less than five large-scale land rushes by white settlers.[17] In the Dakota Territory, the Great Sioux Reservation agreed to sell nine million acres to the federal government. The numbers tell the tale. In 1887, when the Dawes Act was passed, Indian reservations totaled approximately 138 million acres. By 1935, this number had dropped to fifty-two million.

THE FIRST PUBLIC LAND POLICY?

By the end of the nineteenth century, federal land laws and grant programs resulted in the transfer of over one billion acres of the public domain into private hands. As detailed in table 2.1, over 303 million acres were auctioned, sold, or preempted, while another 287 million acres were transferred through the various Homestead Acts. Over sixty-one million acres were given as military land bounties, and ninety-four million acres

Table 2.1. Disposition of the Public Domain, 1781 to 2011

Type of Disposition	Acres
Unclassified	303,500,000
Grants or Sales via Homestead Acts	287,500,000
Grants to States	
Common schools	77,630,000
Swampland reclamation	64,920,000
Railroad construction	37,130,000
Canals and rivers	6,100,000
Wagon roads	3,400,000
Miscellaneous and unclassified	139,300,000
Grants to Railroad Corporations	94,400,000
Military Bounties to Veterans	61,000,000
Timber and Stone Act	13,900,000
Timber Culture Act	10,900,000
State of Alaska and ANCSA	
State conveyances	99,100,000
Native conveyances	43,700,000
Total	1,242,480,000

Source: Based on data from the U.S. Department of the Interior, Bureau of Land Management, *Public Lands Statistics 2011, vol. 196* (Washington, DC: GPO, 2012).

were granted directly to the railroad companies (with another thirty-seven million acres granted to them via the states). The Timber and Stone Act accounted for 13.9 million acres privatized, the Timber Culture Act 10.9 million, and the Desert Land Act another 10.7 million acres more.

The overwhelming "success" of these programs lay in the numerous and diverse means they provided for land to be sold below market value, as well as the ample opportunities they offered for speculation and fraud due to lax enforcement. According to the 1890 census, for the first time in American history, there was no longer a frontier zone as defined by population density. The entire continent was now officially "settled." For some commentators, there no longer remained any place where one might find uncharted, untouched, or otherwise "wild" lands in the United States. At the 1893 World's Fair in Chicago, Frederick Jackson Turner delivered his famous pronouncement that the closing of the frontier marked a new chapter in American history.[18] In terms of privatizing the public domain, the General Lands Office, which oversaw both the lawful exercise and rampant abuse of the land disposal laws, grants, and incentive programs, had done its job well.

Arguably, the disposal of the public domain was the first national environmental policy in the United States. From the nation's founding to the late nineteenth century (and beyond in some cases), federal laws, programs, and regulatory agencies worked feverishly to transfer the national commons into private ownership. And just like the story of public land acquisition, the mammoth effort to dispose and distribute the public domain was interwoven with significant conceptual assumptions that continue to shape our understandings of public lands today.

Prominent among these is the ideal of private property. The influence of this core value in American life cannot be overstated. As we have seen, in roughly the first hundred years of American history, every effort was made to convey the public domain into private hands, whether to small, independent homesteaders or large corporate entities. The primacy of this idea is not surprising given the historical context within which it emerged—namely, the restrictions on land ownership in Europe. Codified in the earliest land ordinances and in the Constitution itself, the right to private property remains one of the most powerful, influential, and definitive values in American society. As we shall see, the acquisition and development of private land was deemed not only economically desirable but also morally righteous and patriotic.

Yet the conceptual significance of private property is much more than the fact that it stands counterpoint to the idea of public land. Rather, it is the way of seeing nature that it imparts—viewing land, water, flora, and fauna as *commodities*—that renders such profound influence. By definition, commodities are objects or goods that can be bought and sold. They can be modified, "improved," divided, reapportioned, transferred, or destroyed at will. Moreover, each of these actions may be taken without consequence to the owner. Commodities are valued according to their utility to consumers, as measured in the marketplace, or in terms of personal sustenance or other preferences.

Conceptualizing land and nature in this way helps explain the ready adoption of the rectilinear survey system as the preferred means of selling and distributing the public domain. Viewed as a commodity, land can be abstracted into equally apportioned units imbued with standardized monetary value, regardless of the diverse environmental realities existing "on the ground."

However, this perspective also has limits that blind us from seeing nature in other ways. It runs counter, for example, to an ecological perspective in which land (or nature) is understood and valued for its role in larger environmental systems (e.g., food chains, nutrient cycles, watersheds, etc.), whose spatial scales and irregular boundaries defy standardized demarcations. A commodity viewpoint also tends to ignore interlinkages in nature. Whereas commodities may be extracted, processed, packaged, and transported without necessary repercussions to the "bottom line," an ecological understanding emphasizes the connections within and between organisms in a system. It acknowledges that changes made in one part of a system necessarily alter other parts of the system and vice versa, often in unintended ways.

But a purely economic perspective ignores all of this. The complexity and intricate connectivity of ecological systems elude easy capture by the rather blunt economic calculus of commoditization. Nonetheless, the legacy of conceptualizing public lands as commodities is profound and relevant to early conservation efforts in at least two ways. First, it provided a lens through which all early conservation proposals were scrutinized. If a portion of the public domain slated for protection was found to have commercial worth, political resistance was often fierce and frequently insurmountable. Second, it meant that even when land was determined as suitable for conservation, frequently its systemic qualities continued to be ignored. For decades, federal land managers treated environmental resources like commodities as a collection of individual pieces of nature that could be dissembled and modified (or managed) without regard to other parts. Such thinking and its unintended consequences often hampered early conservation efforts seeking to manage for single, desirable species or resources.

In sum, whereas the *acquisition* of the public domain corresponded with the conceptualization of public lands as abstract, unpeopled, and pristine landscapes, subject to and corresponding with the birth of the United States, the *disposal* of public lands underscored their conceptualization as commodities for private ownership. And as we'll see in the next chapter, the legacy of these assumptions rendered significant impacts on subsequent efforts to begin managing and setting these lands aside for future generations.

3

A Public Land
System Emerges

The air, the water and ground are free gifts to man and no one has the power to portion them out in parcels. Man must drink and breathe and walk and therefore each man has a right to his share of each.

—James Fenimore Cooper[1]

The year 1872 stands as an extraordinary turning point in the history of America's public lands. It marked the passage of two major federal laws, each reflecting diametrically opposed ways of conceptualizing and responding to the question of what to do with the public domain. Together, these laws highlighted the uncertain and often contradictory relationship between Americans and their public lands in the late nineteenth century. On May 10, President Ulysses S. Grant signed the General Mining Act, the most far-reaching effort to date to commoditize federal lands by formally opening them to mining activity and private ownership. Even today, the law continues to prioritize hard rock mining as the "highest and best use" of public lands vis-à-vis other land use activities and provides the primary legal vehicle for their privatization.

Yet less than three months earlier, on March 1, the president had signed another bill into law. By questioning the idea of nature-as-commodity, this new law appeared to directly challenge the conceptual premise of the General Mining Act. For the first time in history, an enormous portion of the public domain—some 2.2 million acres—was withheld from private development on the rationale that it held aesthetic and inherent value. According to the law, this land could never be sold, settled, logged, or mined. Rather, it was to be preserved in perpetuity as a "public park or pleasuring ground for the benefit and enjoyment of the people." These words refer to the 1872 creation of Yellowstone, the world's first national park.[2]

These two 1872 laws stand as exclamation points that both confirmed (General Mining Act) and confronted (Yellowstone National Park Act) nearly one hundred years of policies and programs designed to sell off the public domain. Their passage created a legal framework that set contrasting American conceptions of nature and the public domain on a collision course—one that would culminate again and again in forceful ideological and political clashes over the future of America's public lands.

But the ideological fragmentation did not end there. In 1891, it reached new levels with passage of the General Revision Act. The primary purpose of this law was to stem the corruption and exploitive land speculation taking place on public lands by repealing the 1873 Timber Culture Act and tightening accountability under the 1876 Desert Land Act. However, in the final minutes of debate, the bill was quietly amended with a rider granting power to the president of the United States to set aside portions of the public domain to create the nation's first forest reserves. Whereas previous laws creating national parks were considered singular cases, this new executive power allowed for the continuous withdrawal of public lands from settlement at the president's discretion. By providing the blueprint for an intentional *system* of publicly owned and managed lands, this action signaled another clear challenge to the dominant view of nature-as-commodity. No longer would the "public" status of public lands be restricted to merely a temporary phase on the road to eventual privatization. Instead, for at least some of these lands, the label now evoked longevity, even permanence, as they were retained in public ownership as part of a new system of national forest reserves. To be sure, the 1891 law did not spell the end of efforts to privatize the public domain. The General Mining Law, various Homestead Acts, and other incentive programs were still in force. And perhaps most importantly, the ideas of nature-as-commodity and as something "unpeopled, eternal and pristine" continued to wield their considerable influence in public debates. But now, for the first time, alongside laws designed to "dispose" of the public domain, precedent-setting laws existed that allowed for the permanent retention and management of public lands for the benefit of all Americans.

TRAGEDY OF THE NATIONAL COMMONS

These watershed events beg the question of what might account for the fracturing of over one hundred years of continuous policies and programs designed to sell off the public domain. Clearly, the drive for land, resources, and economic growth did not suddenly dissipate in 1872 and again in 1891. So what explains the change? In fact, the answer differs considerably depending on the *type* of federal land in question: national parks, forests, or something else? The individual stories—including the rationales, principal characters, contexts, and subtexts—are all unique and explored in later chapters.

Nonetheless, we can still identify a much broader undercurrent of change in the evolution of environmental conservation as a national value. While the emergence of conservation stems from many sources, many environmental historians view its rise to prominence as a response to crisis: the so-called tragedy of the commons (see textbox 3.1). For our purposes, the "commons" is the *national* commons as represented by the public domain. The "tragedy," however, was not a singular event but rather a series of tragedies or actions resulting in environmental degradation, each fueled by

---

TEXTBOX 3.1.
The Tragedy of the Commons

Perhaps the most famous example of the tragedy of the commons derives from a 1968 paper written by Garrett Hardin in the journal *Science*. In this scenario, a commonly owned pasture provides the incentives for its own demise through a simple economic equation. Since each livestock owner receives 100 percent of the benefits from using the pasture but shares equally in the costs of doing so (i.e., for management and upkeep), it is in each individual's economic self-interest to put as many animals on the pasture as possible. As long as each participant puts a limited number of animals on the pasture, the system works. But if they act according to their own short-term personal interests and place additional numbers of livestock on the land, the pasture is soon destroyed by overuse. Uncertainty and fear that one's neighbors will act first drives each livestock owner to the same unfortunate conclusion.

While some have argued that the primary lesson to be drawn from the tragedy is that all forms of common resources (including public lands) should be privatized, others have drawn quite different conclusions. According to scholars such as Elinor Ostrom, the underlying cause of the dilemma is the absence of properly defined and enforced restraints on resource use, not the mere fact that they are commonly owned or managed. And as we have seen in the case of public lands, both the regulatory framework and its enforcement were extremely weak during much of the nineteenth century. In some cases, federal land laws opened the door to new forms of corruption and environmental abuse. Combined with the population growth and technological advances described earlier, their impact on the public domain was both dramatic and widespread.

a complex array of social and economic forces.[3] Rapid population growth abetted by millions of European immigrants, technological advances and market expansions made possible by the Industrial Revolution, and the excesses of a relatively unfettered market economy all contributed to unanticipated social and ecological change across the North American landscape. The rampant use and exploitation of environmental resources led to pollution and pending shortages of clean water, timber, soil, and wildlife that ultimately threatened the health and well-being of American society.

The numbers are telling. In 1783, at the end of the Revolutionary War, the population of the United States was approximately 3.25 million. By 1840, that number had grown to approximately seventeen million people. By 1890, the same year the frontier was declared "closed," the population had exploded to nearly sixty-three million. During that same fifty-year period (1840 to 1890), some twenty-four million immigrants arrived in the United States. Throughout the nineteenth century, annual population growth ranged between 21 percent and 38 percent. While many flocked westward, others joined growing urban centers along the Atlantic Coast and in the Midwest, stretching from Chicago to St. Louis and down to New Orleans. In 1860, New York City became the first American city to boast a population of one million people.

Coinciding with this massive population growth were newly emerging technologies and economic opportunities, all part of the burgeoning Industrial Revolution in North America. The development of late-eighteenth-century inventions such as the steam engine and cotton gin, followed by new farming technologies such as John Deere's steel plow, facilitated the cultivation of agricultural lands across the Midwest. The transportation sector witnessed similar growth and change. As noted in chapter 2, federal and state support for the building of roads, canals, and railroads boomed in the mid-nineteenth century, leading to the first transcontinental line in 1869. The railroads provided western settlers and investors with ready access to urban markets for their agricultural produce, timber, and minerals. Taken together, these advances provided not only incentives but also the means to extract unprecedented quantities of raw materials as inputs for industrial production while creating equally vast amounts of waste.

The impetus for this rush to transform and cultivate the public domain derived from multiple sources. Between the lofty political ideals of Jefferson's agrarian democracy, expressed as a nation of small, independent farmers, and the grounded pragmatism of economic gain derived from the large-scale development of public lands and resources as espoused by the Federalists, stood a third rationale rooted in moral and spiritual beliefs. According to historian Roderick Nash, this moral mandate to conquer, subdue, and otherwise "civilize" nature stemmed from the Judeo-Christian traditions of many early Euro-American colonists.[4] Within settlers' diaries, Nash documents numerous references to God's biblical commandment for humans to subdue nature, to create a tamed and bountiful garden from the unruly wilderness in both a metaphorical and material sense. Over time, Nash argues, this moral and religious righteousness took on a nationalistic tone through the guise of Manifest Destiny.

In addition, scholars often point out that in the early years of westward expansion, settlers interpreted the sheer size of the North American continent as tantamount to an infinite supply of resources. The notion that humans could radically alter, much less threaten, the existence of wilderness on such a vast scale did not seem plausible. Finally, add to this mix of political, economic, and moral incentives a lack of regulatory enforcement over the sale and use of lands and resources on the public domain, and the recipe for disaster was complete. With the necessary ingredients thus assembled, conditions were ripe for the perfect environmental storm.

Given such a powerful mix of causal forces, how did specific tragedies actually play out in the United States? The number and diversity of examples are staggering, but among them, the near extinction of the American bison is particularly poignant. Furs were among the earliest commodity resources actively sought and exported to European markets from North America. Beaver felt hats and bison robes were popular consumer goods. Consequently, by the mid-1800s, many fur-bearing animals began to disappear from their eastern habitats. But out on the Great Plains, bison continued to flourish in two great herds: one in the North and one in the South, split by the transcontinental railroad. By some estimates, bison numbered between thirty and sixty million prior to European settlement. However, from the 1840s to the early 1880s hunting decimated the great herds so that by 1884 only 325 animals remained in the United States.[5]

Because the market for bison robes required thick fleece, the hunting season was generally restricted to the winter months when the animals' fur was the most copious. However, by the 1870s, new tanning processes created additional uses for bison skins in a variety of leather-based products. No longer limited by the demand for heavy winterized furs, the hunting season expanded year-round. Meanwhile, the transcontinental railroads provided new transportation links between the herds and eastern markets. The potential profits rendered by these changes lured professional hunters to the plains. Between 1872 and 1874, they killed over four million bison, while Native Americans took another 1.2 million.[6] By the following year, the southern herd was largely extinct. The northern herd lasted a few more years until meeting its demise with the completion of the Northern Pacific Railroad. In 1880, the line reached Bismarck, North Dakota, and three years later, except for a few straggling survivors near Yellowstone and elsewhere, the northern herd followed its southern neighbors into oblivion.

Another powerful illustration of the nineteenth-century tragedy of the commons can be found in the story of the eastern forests. Ever since the colonial period, timber has been a highly prized commodity, used for fuel wood, construction, furniture, railroad ties, mining structures, and charcoal in iron furnaces. Logging companies, operating both legally and illegally, clear-cut much of the hardwood forests of Appalachia as well as the white pine woodlands of the Northeast and Great Lakes region. A common practice was "high grading": cutting down most trees in an area but then taking only the most valuable logs to market. Making no effort to replant or allow for natural regeneration, loggers would leave huge piles of wood slash and debris in their wake. Once dry, this waste material became fuel for massive forest fires.

The most destructive fire in U.S. history resulted precisely from these conditions. In 1871, the Peshtigo Fire in northeastern Wisconsin covered 1.28 million acres and killed over 1,500 people.[7] With the tree cover and vegetation burned away, top soil quickly eroded into rivers and streams, destroying or damaging water resources and creating conditions for subsequent flooding. According to one estimate, such logging methods and associated wildfires resulted in the clearing or degrading of over eighty million acres of forests east of the Mississippi.[8]

Similar scenes of environmental devastation could be found across the country, from the gold and silver mining districts in California and Colorado to the smoke-filled skies of Chicago, New York, and other expanding urban areas along the eastern seaboard. Barren stretches of denuded landscapes replaced the once-thick forests of the Great Lakes region, while the high-grass prairies of the eastern Great Plains, once filled with the great bison herds, were now overgrazed with massive herds of domestic cattle. Amid these dramatic images, new voices emerged that questioned the logic of unrestrained privatization and industrial "progress." Together they posed the question, paraphrased from Aldo Leopold: What if we stopped viewing nature—and hence, the public domain—in purely economic terms?

FROM CRISIS TO CONSERVATION

Calls for change took many forms and emanated from a wide variety of sources. Nonetheless, they coalesced around the shared notion that America's natural resources were under threat. The only reasonable response was a new way of conceptualizing, valuing, and managing the public domain. Though situated at the margins of popular opinion, these voices succeeded in laying the groundwork for the later achievements of the conservation movement at the turn of the century.

One of the earliest advocates for change was the frontier artist George Catlin. Famous for his paintings of Native Americans and western landscapes, Catlin wrote a memoir in 1832 in which he advocated setting aside a portion of the public domain as a national park.[9] Specifically, he envisioned a preserve running north and south along the western Great Plains to save from extinction both the American bison and the Plains tribes who depended upon them for their livelihood.

In following decades, a growing chorus of supporters joined Catlin. The American Romantic movement served as a particularly rich source. Like its European predecessor, American Romanticism did not enjoy a mass political following. Rather, it was expressed through the arts: literature, poetry, and the landscape paintings of the Hudson River School. Romantics challenged the dominant social conventions of early industrial society, advocating emotion, spontaneity, intuition, and individual freedom over the cold economic rationality of modern urban life. Premised on nostalgia for a pristine past, the celebration of aesthetic beauty and the recognition of inherent value in wild nature, Romantics supported the preservation of nature for its own sake.

Transcendentalists, such as Ralph Waldo Emerson, went a step further, flipping the Euro-American notion of an immoral wilderness on its head. To commune with God, he argued, one should seek out the sublime and untrammeled forms of wild nature, not "conquer," civilize, or otherwise transform it into a pastoral garden. Corruption resided in human civilization, especially in its urban industrialized form. For Emerson, the search for honesty and purity necessarily led one to the wilderness, where nature served as a conduit through which God expressed spiritual truths. In 1851, Emerson's friend and protégé, Henry David Thoreau, echoed Catlin's earlier call for the preservation of wild nature in national parks, famously declaring that "in wildness is the preservation of the world."

However, it was not until writers such as John Burroughs and John Muir took up the preservationist cause that these ideas gained a following large enough to constitute a potent national political force. In articles in newspapers and magazines, Muir's words painted beautiful pictures of wild nature coupled with passionate pleas for its protection. His efforts contributed to the creation of additional national parks on the model of Yellowstone, including Yosemite, Mount Rainier, and Sequoia National Parks. In addition to his writings, Muir founded the Sierra Club in 1892, which has worked to garner public support for the preservation of nature long after Muir's death.

Other calls for change were premised less in spiritual truths than in the pragmatic concerns of resource depletion. In 1864, George Perkins Marsh published *Man and Nature: Or, Physical Geography as Modified by Human Action*. The book was highly influential, documenting how the historical decline of past civilizations was linked to the degradation of their natural resource base. In so doing, Marsh underscored the link between nature and human society, demonstrating how the abuse and overuse of environmental resources could render deleterious ecological and social effects. Avoiding this demise required careful, science-based regulatory limits on resource use or, in other words, environmental conservation.

Marsh's concerns were echoed by a number of his contemporaries. In 1878, John Wesley Powell, a one-armed Civil War hero made famous by his exploratory expedition down the Colorado River, issued his *Report on the Lands of the Arid Region of the United States* to Congress. The report promoted scientifically informed land management in order to facilitate the settlement of the desert Southwest. Recognizing the fundamental role of water resources in the region, he proposed large-scale, government-funded irrigation projects (ultimately adopted) and administrative jurisdictions based on watershed boundaries rather than arbitrary rectilinear survey lines (largely ignored).

Within the federal government, concerns over resource degradation and waste were also on the rise. Carl Schurz, Secretary of the Interior from 1877 to 1881, was one of the first federal officials to address rampant land fraud on the public domain by promoting federal regulation. Meanwhile, the formation of new nongovernmental organizations, such as the Boone and Crockett Club—established in 1887 by the likes of George Bird Grinnell (editor of *Field and Stream* magazine) and a young Theodore Roosevelt—mobilized new constituencies of sportsmen in favor of maintaining viable populations of wildlife and, by extension, their habitats. Taken together, this diverse group of Romantic idealists, naturalists, politicians, sports enthusiasts, and early scientific conservationists drummed up sufficient public support for the creation of a federal system of national parks, forests, and wildlife refuges.

This coalition was anything but ideologically or politically monolithic. Rather, it was propelled by diverse and even conflicting motives and conceptions of the public domain. Among the most prominent fissures was the distinction between conservation and preservation. Whereas preservationists upheld the idea of inherent value in nature and advocated that some wild places be left untouched and undeveloped, conservationists continued to support the goal of resource development. For conservationists, human utilization was still the overriding purpose of the public domain. This priority was never in question. Instead, they took issue with the *manner* in which it was achieved. In particular, they sought to replace the profit motive for resource development with scientific management, applied in such a way as to maximize the efficient use of resources over time for the benefit of the common good.

In the competition of ideas for an alternative approach to public land policy, the idealism of Romantic preservation generally lost out to the pragmatism of environmental conservation. While many view preservationist ideals as the driving force behind national parks and wilderness areas (see chapters 4 and 8), most other types of federal lands are considered products of conservation (e.g., national forests, wildlife refuges, and so on). Strongly associated with the presidential administration of Theodore Roosevelt, progressive conservation, as it came to be known, was part of a much larger platform of initiatives that reached well beyond environmental concerns. Reforms of the Progressive Era (approximately 1890 to 1920) addressed issues ranging from public education and women's suffrage to industrial labor laws, monopoly "trust busting," and prohibition. But in terms of public lands and resources, the concept of progressive conservation achieved its fullest expression—and received its most ardent support—in the hands of Gifford Pinchot.

As first chief of the U.S. Forest Service and confidant of President Teddy Roosevelt, Pinchot was extremely influential. He not only shaped the principles of conservation but put them into action. According to Pinchot, conservation stood first and foremost for the development of natural resources. They should be used, not preserved. However, for Pinchot, resource development was not designed to line the pockets of the wealthy. Rather, it should be practiced in such a way as to increase efficiency, reduce waste, and provide benefits to all Americans. To achieve this goal, Pinchot needed a corps of elite, scientifically trained resource managers that could be relied upon to resist the temptations of abusing natural resources for short-term financial gain. In 1905, the USDA Forest Service was born to oversee the newly created national forest reserves. The famous phrase "the greatest good, for the greatest number in the long run" became a mantra for the Forest Service, and it was reflected in other Progressive Era agencies established to oversee lands and resources in the public domain.

During this period, the government also established the Bureau of Reclamation in 1902 to carry out large-scale irrigation projects in the arid West. And in 1916, Congress rearranged the piecemeal collection of national parks into a system overseen by the new National Park Service located within the Department of the Interior. Other types of public lands, such as national wildlife refuges and wilderness areas, would have to wait several more decades to receive their own dedicated federal management agencies and foundational legislative action.

Though some question the sharp distinction between preservation and conservation in terms of actual management practices and outcomes (is managing for tourism really a preservationist ideal?), the tensions between these two viewpoints and their respective advocates were real enough. The famous debate between John Muir and Gifford Pinchot over damming the Hetch Hetchy Valley in Yosemite National Park is a case in point (see chapter 4). These unresolved tensions, along with the challenges posed by the legacy of dominant eighteenth- and nineteenth-century conceptual assumptions, would shape public land discussions for years to come. In the twentieth century, many additional strains of environmentalism would evolve that built upon these distinctions, at times complementing, and at times competing with, one another. But despite all of this, the institutional blueprint was now set for the development of a diverse system of publicly owned and managed lands and resources.

BUILDING THE PUBLIC LAND SYSTEM

As noted earlier, the 1872 creation of Yellowstone National Park stands as the world's first continuously existing national park. However, it did not represent the first federal action to set aside portions of the public domain for conservation purposes. As discussed in the following chapter, Yosemite Valley and the Mariposa Grove of sequoia trees were both set aside from settlement in a bill signed by President Abraham Lincoln in 1864. However, the law gave management of the two parks to the state of California. They would not become part of the national park system until 1906. An even earlier example occurred in 1832, when Congress set aside four 640-acre sections to create the Arkansas Hot Springs Reservation. But settlers generally ignored this designation until an 1877 lawsuit and the formal establishment of the land as a national park in 1921. In fact, one can reach back all the way to the colonial period to find numerous additional cases. At different times, certain resources such as timber for ship masts and deposits of lead and salt received nominal governmental protection. But the difference is that none of these declarations were long lived. What makes Yellowstone distinctive is that it remains the oldest continuously protected national park in the United States and, indeed, the world. It therefore represents the "first piece" of what would later become a national public land *system*.

After Yellowstone, generally speaking, the growth of America's public land system took place during four distinct periods. These include the Progressive Era, from the 1890s to the early 1920s; the New Deal Era of the late 1930s and early 1940s; the "Environmental Decade" of the 1970s; and finally, what I term the "Pendulum Years," from the 1980s to the present. This final period reflects the full politicization of public land issues along partisan lines, tracking the ebb and flow of environmental protection and deregulation in concert with changes in the national political landscape. However, it also marks the rise of ecosystem management and collaborative conservation as alternative strategies for moving beyond partisan gridlock.

In each of these four periods, the impetus for change derives from episodes of social and ecological upheaval in American society. If one attempts to map this out, a distinctive three-stage pattern readily emerges (figure 3.1). The first stage is characterized by relatively unfettered development of lands and resources, fueled in part by the rise of new technologies, products, industries, and population growth. Over time, the various and often unanticipated impacts of these dynamics culminate in a second stage of social, economic, and ecological crises. For public lands and resources, these crises often translate as new forms of air and water pollution, resource depletion, or habitat loss, all of which have significant social causes and consequences. The regulatory response to these problems constitutes a third stage, one that typically expands and strengthens environmental protections. In this stage, new leaders emerge to mobilize public support for governmental action: setting aside new portions of the public domain, establishing new management agencies, and passing new land use laws. After a time, deregulatory sentiment finds a new audience, new rounds of innovative technologies emerge, and the cycle repeats.

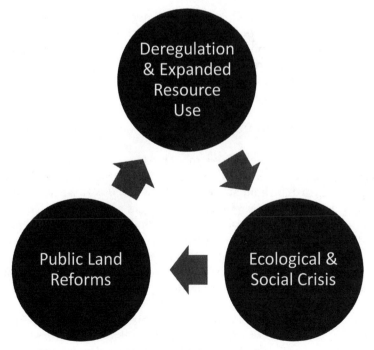

Figure 3.1. A Three-Stage Cycle of Public Land Policy

Thus, after several decades of industrial expansion and unregulated resource use, the social and environmental crises of the late 1800s provided
the catalyst for the burgeoning environmental conservation movement
in the United States, dually expressed in the ideals of American Romanticism and the policy reforms of the Progressive Era. These crises and
responses provided the initial framework for America's system of public
land conservation and management. National parks, forests, and wildlife
refuges came into being during this time, along with the federal agencies
charged with their care.

However, these reforms faced a deregulatory reaction after World
War I. Claiming that many Progressive Era changes had gone too far,
the so-called Roaring Twenties marked a period of rampant government
corruption, organized crime, weak regulatory enforcement, and relatively
unencumbered resource extraction and development. By the end of the
decade, conditions were once again ripe for a new set of crises in American society, both social (the Great Depression) and ecological (most dramatically illustrated by the Dust Bowl). These events, in turn, would lead
to a new round of expansion for America's public land system.

New Deal Reforms

In the wake of crisis, several new voices emerged in the 1930s and 1940s
to champion the cause of public land protection. Writing in *Harper's
Magazine* and other popular outlets, Bernard DeVoto made a persuasive
case for protecting the remaining unhomesteaded portions of the public
domain against privatization. As western politicians and livestock interests sought to transfer ownership of federal grazing lands (and other
forms of public lands) into state and private hands, DeVoto's cogent
defense of the public land system helped to squelch congressional support for such a plan.

Meanwhile, former U.S. forest ranger Aldo Leopold argued for new
approaches to soil conservation, wilderness protection, and wildlife
management. In his seminal work, *A Sand County Almanac* (published
posthumously), Leopold presented an eloquent appeal to farmers, sportsmen, and private landowners to develop a "land ethic": an understanding
of, and appreciation for, nature that would evoke conservation measures
without waiting for government handouts or mandates. In 1935, Leopold, Robert Marshall, Benton MacKaye, Robert Sterling Yard, and others
formed the Wilderness Society to protect lands that "yet remain free from
mechanical sights, sounds and smells."

At the federal level, President Franklin Delano Roosevelt responded to
both the economic and ecological crises of the day with his famous New
Deal programs. Similar to earlier Progressive Era reforms initiated by his

distant cousin Teddy Roosevelt, FDR's programs established new federal agencies and land use regulations. For public lands, these included the U.S. Fish and Wildlife Service to oversee the system of national wildlife refuges; the first efforts to regulate grazing on the public domain via the 1934 Taylor Grazing Act; and the creation of both the Civilian Conservation Corps and the Soil Conservation Service (later renamed the Natural Resource Conservation Service). The Truman administration continued this trend with the establishment of the Bureau of Land Management in 1946.

In the post–World War II era, however, new technological advancements, economic development, and population growth brought unforeseen changes to the public domain. One of the most substantial was triggered by the growth of the middle class, whose new prosperity allowed the purchase of new homes, automobiles, and consumer goods in unprecedented numbers. With expanded leisure time and heightened mobility (enhanced by the 1955 completion of the interstate highway system), visitation to public lands for recreational purposes skyrocketed. For the national parks and wildlife refuge system, this translated into an enlarged constituency, new car-friendly improvements, and an expanded capacity to accommodate visitors. For the national forests and BLM lands, it meant a diversification of their traditional constituencies of timber, mining, and livestock industries. New management priorities, such as the provision of campgrounds, roads for recreational use, wildlife protection, and aesthetics, increasingly vied for attention with long-established commercial concerns for resource extraction.

Coupled with advances in the communication, transportation, and construction industries, postwar population growth also spurred the expansion of cities and suburbs. New production processes and advances in chemical industries led to an outpouring of new consumer products, increased economic productivity, and corresponding increases in the extraction of raw materials needed for production. Industrial pollution into the air and waterways surged at rates not seen since the Industrial Revolution. But by the late 1960s, the rapid postwar economic growth in the United States was beginning to stall, exacerbating a variety of social tensions. The baby boom generation, now attending college in unprecedented numbers, began to question long-standing social norms, fueling conflicts over war in Vietnam, race relations, and women's rights. The United States was embroiled once again in a series of intertwined social and ecological crises.

The "Environmental Decade"

Amid this turbulence, headlines across the country documented a series of new environmental disasters. Scientists warned that the national symbol of the United States—the bald eagle—was in danger of extinction. Spills from offshore oil rigs sent huge mounds of oil tar to California beaches, smothering wildlife and threatening tourism. In Ohio, images of the Cuyahoga River in flames, so full of pollutants that it actually ignited, flashed across the nightly news. And through it all, the visibly polluted skylines of America's major cities provided a fitting, if ominous, backdrop.

In the midst of these events, John Muir's writings found a new audience and relevancy, as did Aldo Leopold's *A Sand County Almanac*. Meanwhile, a new generation of environmental writers such as Barry Commoner, E. F. Schumacher, and Edward Abbey took up their pens. Undoubtedly, the most influential of these was Rachel Carson. In her 1961 book, *Silent Spring*, Carson offered a compelling, science-based documentation of the problem of environmental pollution in the modern age. Giving special attention to the problem of DDT, she carefully outlined how the toxin could spread through the ecosystem—into waterways, airways, and food chains—to detrimentally affect both wildlife and humans. By effectively illustrating the link between human and environmental health, *Silent Spring* significantly expanded the constituency for environmental protection. No longer limited to those interested solely in wild lands or national parks, environmental advocates now included activists focused on living conditions in cities, suburbs, and rural areas. Carson's work, coupled with the string of current national and global events, helped to generate a groundswell of support for a new grassroots movement: modern environmentalism.

According to environmental policy expert Robert Gottlieb, one key to the early success of the environmental movement was the fact that it was viewed as a politically neutral issue.[10] Given the harsh partisan divide on so many other matters of the day, from civil rights to the Vietnam War, politicians were keen to find areas of common ground, and environmentalism appeared to fit the bill. Who could be against the idea of clean air or water? Democrats and Republicans vied with each other for the mantle of "environmental protector," producing bipartisan support for a wide range of environmental measures. Under these rare political conditions, the so-called Environmental Decade was born.

In actuality, this "decade" stretched from to the mid-1960s to the end of the Carter administration. The Clean Air Acts of 1963 and 1970, the Clean Water Acts of 1972 and 1977, the Endangered Species Act (ESA) of 1973, and the Resource Conservation and Recovery Act (RCRA) of 1976 are just a few of the foundational laws passed during this time. In many cases, these laws represented the very first national-level regulatory efforts to address various environmental problems. And with the 1970 creation of the Environmental Protection Agency by President Richard Nixon, there was now a federal agency with the mandate to enforce them.

For public lands, this heady period was no less significant, marking the passage of laws allowing the designation of new types of public lands, including the Wilderness Act of 1964, the National Wild and Scenic Rivers Act of 1968, and the National Trails Act of 1968. Meanwhile, the National Environmental Policy Act (NEPA) of 1969 mandated a precautionary rather than a reactionary approach to public land management with the requirement of environmental impact statements (EISs). These statements sought to identify the social and ecological effects of proposed and alternative actions on federal lands before a new management decision could be made.

Many of these new environmental laws not only mandated new regulatory procedures but also tested some of the dominant conceptual assumptions underlying public lands and resources. For example, the notion of absolute national sovereignty over public lands was softened with new "citizen suit provisions" contained in several laws, including the 1970 Clean Air Act and 1972 Clean Water Act.[11] Now, U.S. citizens could challenge decisions of federal resource managers in court if they could show that the managers had not followed legally defined standards or procedures. Moreover, requirements in NEPA, the 1976 National Forest Management Act (NFMA), and the 1976 Federal Land Protection and Management Act (FLPMA) mandated public participation not only at the end of the decision-making process but at the beginning as well. In this way, the public began to play a more significant role in the formulation of resource management plans.

In a similar vein, several of the new laws began to question, at least indirectly, the concept of nature-as-commodity. Language in both NEPA and the ESA explicitly blamed unplanned economic development for creating modern environmental problems. Instead they proposed consideration of other ways of valuing land and nature, not solely as commodities, but as elements with inherent value. The 1973 Endangered Species Act went even further, calling for the protection of critical habitats of species listed as either threatened or endangered with extinction on both public and private lands, without requiring assessment of associated economic costs (at least until a 1978 amendment). Passage of the 1964 Wilderness Act and the Sierra Club's successful campaigns to halt proposed dam projects in Dinosaur National Monument (1956) and Grand Canyon National Park (1968) further underscored these points.

Finally, two more major laws capped off the Environmental Decade, both passed in 1980. The Superfund Law (formally known as the Comprehensive Environmental Response, Compensation and Liability Act) dealt with cleaning up toxic and hazardous waste spills in the wake of the Love Canal debacle. However, it was the second law, the Alaska National Interest Lands and Conservation Act (ANILCA), that fundamentally transformed the public land system in the United States. This law expanded the national park, national forest, and wilderness and national wildlife refuge systems by more than 79.5 million acres. In one fell swoop, the Act doubled the amount of land in the national park and national wildlife refuge systems, and it tripled the size of the national wilderness preservation system.

One conceptually important aspect of ANILCA was its departure from the notion of the public domain as "empty" or uninhabited with regard to indigenous peoples. Elsewhere in the lower forty-eight states, Native Americans had often been removed to reservations prior to settlement or public land designation. However, in Alaska, they still occupied many of these lands and demanded a voice in management decisions. The notion of human communities existing *within* the bounds of a wildlife refuge or national park—common outside of the U.S. experience, but unheard of at home—presaged the role of ecosystem-based approaches to resource management and sustainable development on America's public lands.

The Pendulum Years

From Backlash to Ecosystem Management

The 1980s ushered in another period of deregulation and the returning influence of nature-as-commodity ideals. President Ronald Reagan's first term was marked by his campaign statement, "I am a Sagebrush Rebel." His choices for Secretary of the Interior, James Watt, and Secretary of Agriculture, Anne Burford, reflected these ideals. Each encouraged resource development and privatization of the public domain and worked hard to weaken the new environmental regulations of the previous decade, citing concerns over economic cost.

Yet the 1980s also marked the rapid growth of environmental nongovernmental organizations. The Sierra Club, the Wilderness Society, the Nature Conservancy, and others watched their memberships soar. By mid-decade, the Reagan administration's deregulatory approach to public lands began to falter. Watt and Burford were replaced, and Congress passed laws strengthening, rather than weakening, existing environmental laws. In addition, the 1980s witnessed the growth of a new concept in resource management among federal managers: ecosystem management. By viewing nature as an interconnected system, officials began to consider management actions based on the linkages between wildlife and habitat dynamics as they might occur on a watershed scale. This was a significant departure from conventional management, which tended to focus on a single species or resource within an ecosystem.

Nonetheless, by the end of the decade, the country again appeared poised for a new round of social and environmental crises. With an economic recession looming, a number of ecological events alarmed the public like never before. The 1989 *Exxon Valdez* oil spill in Prince William Sound, Alaska, raised questions over the lack of oversight and general arrogance among large oil companies regarding the environment. Confirmation of "holes" or thinning in the stratospheric ozone layer over Earth's poles led to the international banning of CFCs (chlorofluorocarbons) via the 1987 Montreal Protocol and the 1990 Clean Air Act Amendments in the United States. Growing evidence of climate change marked the first years of the new decade, an issue that gained widespread international recognition at the 1992 Earth Summit in Rio de Janeiro convened by the United Nations. Scientists began to better understand that humankind was not only capable of but had already begun to affect environment change on a truly global scale.

On public lands, debates raged over the clear-cutting of national forests and the practice of "below market cost" grazing fees and timber sales to private ranchers and logging companies, respectively. However, the most famous crisis culminated over the potential listing of the northern spotted owl under the Endangered Species Act. Not since the dam controversies over Hetch Hetchy and Echo Park had national attention focused so intently on a public land issue. The owl's habitat extended throughout the old growth forests in the Pacific Northwest. Thus, an endangered listing meant potentially limiting or closing off these tracts of public land to logging. Appearing on the cover of *Time*, the owl galvanized divisions between environmentalists and resource development interests.[12] More specifically, the battle pitted environmental groups and federal scientists against a coalition of international logging industries, local rural residents, and privatization advocates. The latter successfully characterized the debate as a "jobs versus the environment" issue and called for a deregulatory solution under the guise of "wise use" management (harkening back to one of Pinchot's terms for progressive conservation). Attempting to resolve this polarizing conflict through collaborative means, the newly elected Clinton administration convened the Northwest Timber Summit. Bringing all stakeholders to the table, the Summit laid the groundwork for the 1994 Northwest Forest Plan, and in the process, helped usher in a new form of decision making that has continued to shape American public land management to this day: collaborative conservation.

A Third Way? The Rise of Collaborative Conservation

At least initially, President Bill Clinton's White House victory appeared to signal a shift toward greater environmental regulation in the United States. However, the historic Republican victory in the House of Representatives in 1994 meant that compromise would be the order of the day. While few new environmental laws passed through Congress,[13] the Clinton administration made effective use of executive orders and administrative rule changes to advance their agenda, including the creation of the first Office of Environmental Justice within the EPA and the establishment of the Grand Staircase-Escalante National Monument in southern Utah.

Significantly, polarizing debates over environmental issues were by no means limited to the Pacific Northwest. In the early 1990s, conflict erupted throughout the western United States in rural communities undergoing rapid socioeconomic change. In many places, traditional industries such as logging, ranching, and mining were declining in the face of heightened global competition. At the same time, growth in service-sector industries along with technological advances in telecommuting allowed many people to move or retire to rural environments in record numbers. These so-called New West transformations not only brought new residents into rural communities but new tensions as well. Notably, many of these ex-urban migrants brought with them diverse values and priorities for the management of nearby public lands, values and priorities that often clashed with those of longtime residents. Desires for expanded recreational opportunities, greater protections for endangered species, or the preservation of aesthetic values did not align well with traditional interests in logging, ranching, and mining. As the number of new residents with New West environmental values grew, debates over public lands became increasingly mired in political gridlock. In addition to the Wise Use Movement, a County Supremacy Movement sprang up in some western counties, whereby local government officials declared ownership over all federal lands within their borders.[14] By the mid-1990s, the popular media described the rising number and intensity of conflicts as the latest round of "Unrest in the West."[15]

In response, some residents, federal managers, and local government officials attempted to move past the gridlock by adopting collaborative approaches to public land decision making. One of the first examples to gain national attention was the 1994 Northwest Forest Plan. Seeking to replace the "jobs *versus* environment" argument with the notion of "jobs *and* the environment," the plan offered a comprehensive, ecosystem-wide management strategy to balance the socioeconomic needs of local communities with the regulatory protections needed to preserve the ecological integrity of the region. To reach this middle ground, managers used a collaborative community-based approach, bringing environmentalists, loggers, local residents, and government officials to the table as part of the decision-making process.

Today, environmental conflicts in the Pacific Northwest continue to be complicated by debates over logging, endangered species, and the changing economic fortunes of the region. However, community-based collaborative approaches to resource management have spread across the nation. Grassroots initiatives such as the Quincy Library Group in California, the Applegate Partnership in Oregon, and the Ponderosa Pine Forest Partnership in Colorado represent early examples in forest and watershed management.[16] By the late 1990s, it was estimated that over 97 percent of national forests in the United States had implemented some form of community-based collaboration as part of their management and planning activities.[17]

Collaborations can be "horizontal," between agencies at the same level of government, or "vertical," between different scales of government, including local residents, state agencies, and federal officials. The precise form and structure in any particular collaborative conservation effort varies widely depending on the resource in question, the institutions and local interests involved, and the geographic scale or configuration of the management area. As a result, collaborations can be initiated by grassroots efforts or government agencies. Public participation may be open to all local residents or restricted through a selection process that determines "relevant stakeholders," while end products can range from comprehensive resource management plans to resolutions for site-specific land use conflicts. According to proponents, the approach seeks to increase the quality and quantity of local public involvement in public land management and planning processes.[18] By bringing all interested stakeholders to the table, the approach encourages open dialogue and collaborative learning among and between disparate local land use interest groups and government resource managers. The goal is to establish common ground among participants leading to resource management decisions that balance ecological health with local economic development concerns, to increase local support for the implementation and monitoring of management projects, and to cultivate social cohesion through stronger community-state and intracommunity relations.[19]

Or not. Collaborative conservation is not without its critics. Well aware of the potential it holds to undermine or weaken hard-fought-for national environmental regulatory standards, many leaders of national environmental organizations and federal agencies have been understandably wary of collaborative decision making. By offering local residents unprecedented participation in the crafting or the implementation of management plans, critics fear how initiatives designed to protect environmental resources may be weakened to facilitate the economic concerns of local interests (or those that claim to be).

Nonetheless, the sustained popularity of collaborative conservation by presidential administrations over the past two decades has been nothing short of astonishing. It is one of the few areas of public land management that has received consistent bipartisan support at the national level. Ushered in during the Clinton administration, collaboration received strong support from both the George W. Bush administration *and* the Obama administration.

The staying power of collaborative conservation may stem from the appeal it holds for diverse stakeholders in public land management debates. For environmental scientists, it speaks to the idea of ecosystem management. By bringing all landowners and managers involved in a single watershed together, it makes possible a management plan that addresses the entire system rather than one component part. Moreover, collaboration allows consideration of a key element that is often excluded from resource management plans: the human communities residing within the ecosystem.

For federal managers, collaboration offers a way to include local residents not only in decision making but also in the implementation and monitoring processes that follow. If locals develop a sense of ownership in a management plan, they are more likely to support it and provide much needed voluntary assistance to resource-strapped federal managers. This may also translate into fewer legal challenges to agency decisions.

For local residents, it provides greater access to management processes and a potential way to resolve conflicts by opening dialogue and building a sense of community cohesion. "Old West" residents involved in conventional land use activities (logging, ranching, and mining) may find collaboration provides the best way to still wield influence in the face of economic decline and stronger national regulatory standards. And for residents with interests in environmental protection or recreational opportunities, it allows a mechanism for including local knowledge and tailoring management decisions to unique ecosystem or place-based characteristics.

The New Century

Aside from the issue of collaborative conservation, political support for America's public lands has already completed two "full swings" since the dawning of the twenty-first century. During the 2000s, the George W. Bush administration launched numerous efforts to roll back national environmental regulations in favor of resource development on the public lands. However, like the Clinton administration, the Bush administration soon learned that a divided Congress was not the most effective means of realizing its goals. Legislative attempts to reform laws such as the Endangered Species Act or to open the Alaska National Wildlife Refuge to oil drilling repeatedly fell short. Though the administration achieved a few victories with passage of the Readiness and Range initiatives, the 2005

Energy Act, and the 2003 Healthy Forest Restoration Act, most efforts to roll back public land protections focused on rule changes that did not require congressional approval. In this way, the administration ended U.S. participation as a delegate to the Kyoto Protocol meetings and refused to regulate carbon dioxide under the Clean Air Act (though this was later overturned by the Supreme Court; see below). On public land issues, the Bush administration's most significant impacts included postponing implementation of the Roadless Area Rule for national forests, opening huge tracts of public land for new energy development, and expanding logging activity on national forests, under the logic that it would reduce the threat of wildfire and disease outbreaks according to the Healthy Forest Restoration Act.

One rather unexpected action by President Bush was his declaration of several massive marine national monuments in the Pacific Ocean, including the 140,000-square-mile Papahanaumokuakea Marine National Monument in 2006, and three new monuments near the Mariana Trench in 2009. These decisions stand in stark contrast to the deregulatory environmental policy trajectory pursued by the Bush administration for most of its tenure. While the declarations were not without controversy (see chapter 6), environmentalists heralded them, nonetheless, as a positive step.[20]

With the ascension of the Barack Obama administration in 2009, the political pendulum swung back once again toward support for environmental regulatory protection. The signature piece of public land legislation in his first term was the 2009 Public Land Omnibus Act, the largest land-based protection law in twenty-five years. The law designated over two million acres of public lands as wilderness areas in nine states, established over one thousand miles of wild and scenic rivers, created three new national parks, three new national conservation areas, four national trails, and ten new national heritage areas. With backing from Supreme Court decisions in 2007, Obama also ordered the EPA to begin regulating carbon dioxide emissions, but he has yet to achieve national climate legislation. Most recently, debates over the Keystone Oil Pipeline are testing the president's commitment to alternative energy production, as are environmental campaigns to reform federal grazing fees, reform the General Mining Law, and reduce drilling permits on public lands. While the Obama administration continues to support collaborative conservation, questions remain with regard to how much he will achieve in strengthening or expanding public land protections in an era of continued political division. Professing a commitment to compromise, the president may find that collaborative public land management offers one of the best chances of realizing his vision of a more sustainable future for the public domain.

Part II

AMERICA'S PUBLIC LAND SYSTEM

4

National Parks

Everybody needs beauty as well as bread, places to play in and pray in, where Nature may heal and cheer and give strength to body and soul alike.

—John Muir[1]

Be it enacted that . . . the tract of land near the headwaters of the Yellowstone River . . . is hereby reserved and withdrawn from settlement, occupancy, or sale under the laws of the United States, and dedicated and set apart as a public park or pleasuring ground for the benefit and enjoyment of the people.

—Act Establishing Yellowstone National Park, 1872

Have you ever stood quietly in the early evening to watch the last rays of sunlight glance off the granite walls of Half Dome in Yosemite? Sat on the rim of the Grand Canyon and wondered at the magnificent sheer expanse of space, glowing and shimmering with banded hues of yellow, orange, and red? Felt the rumbling surge, deafening roar, and cool, draping mist beneath Yellowstone's Old Faithful? Or woke at dawn to take the long, winding traverse up the ethereal Going-to–the-Sun Road in Glacier? For those lucky enough to experience them, the very names of these places evoke images of grandeur; vast and dramatic landscape vistas that mix color, scent, wind, heat, and coolness into a rich celebration of nature and memory. But even those who have never set foot in a national park find them deeply woven into the cultural and symbolic fabric of the nation.

Without question, the national parks are the most well known and visited of all types of federal lands in the United States. And this is no accident. Their popularity is due in part to their brilliant natural beauty and, in some cases, profound historical significance, but it also derives from decades of concerted effort by both the national government and private industry to attract and accommodate visitors. This emphasis on recreation has always sat uneasily next to the other major purpose of the national parks: preservation. As the 1916 National Park Service Act states, the parks should be managed in such a way as to leave them "unimpaired for future generations." As places imbued with inherent value, the parks are thus intended for preservation in their original state. Hence, with few exceptions, there is no hunting, no mining, no commercial logging, and no livestock grazing allowed in the parks (activities that continue to varying degrees on most other types of federal lands). Nonetheless, the tension between use and protection remains one of the major challenges for National Park managers.

Today the lands making up the national park system include 401 units, totaling over eighty-four million acres (figure 4.1) across eighteen different designation types.[2] They range in size from the tiny .02-acre Thaddeus Kosciuszko National Memorial in Pennsylvania, to the mammoth Wrangell-St. Elias National Park in Alaska, topping out at over thirteen million acres. And the landscapes differ wildly: from the lofty, snow-covered peaks of Mount McKinley (Denali) to the arid lowlands of Death Valley; and from the lush forestscapes of Acadia in Maine to the sun-drenched sands of Arches National Park in Utah. Beyond areas of natural significance, the park system has evolved to encompass a sundry assortment of places infused with cultural and historical value, including sites such as Ellis Island in New York and the National Mall in Washington, D.C. This expanded system of land types managed by the National Park Service includes national parks, national monuments, national historic sites, national recreation areas, national preserves, national seashores, lakeshores, memorials, battlefields, wild and scenic rivers, historic trails, parkways, and more (see appendix B). Each of these designations carries different levels of environmental and/or historical protection. While the national parks remain distinct within the system, their history is closely intertwined with these other public land designations and the conceptual challenges they evoke.

But of course, it was not always this way. The evolution of the national parks came about in fits and starts. As discussed in chapter 3, it was the culmination of many voices calling for change and the mobilization of diverse constituencies to confront, and at times work within, the dominant conceptions of nature and the public domain at different points in history. It also involved the removal of indigenous peoples who, for millennia,

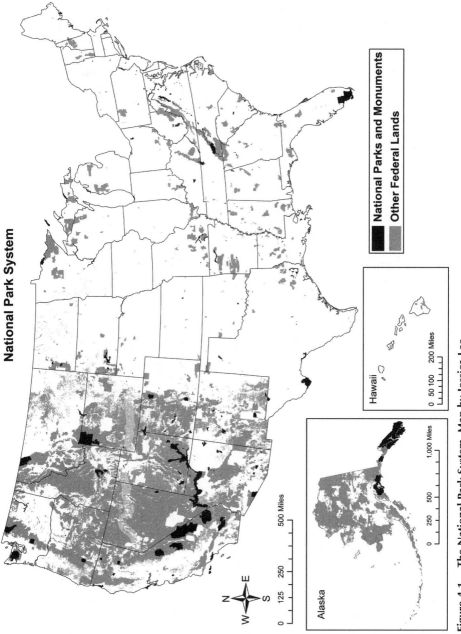

National Park System

National Parks and Monuments
Other Federal Lands

Hawaii

0 50 100   200 Miles

Alaska

0   250   500   1,000 Miles

0 125 250   500 Miles

N
W   E
S

Figure 4.1.   The National Park System. Map by Jessica Lee.

had called these places home.[3] The notion of public lands as unpeopled, eternal and pristine, of nature-as-commodity, and even the strain between preservationist and conservationist ideals all played significant if unacknowledged roles in these struggles. Still, each stage of development in the system—and indeed, the creation of each park—was highly contextual, reflecting the times and issues of the day. The question of our relationship with nature has always been intertwined with broader issues of public discourse and debate, and this is no less evident in the story of the world's very first national park: Yellowstone.

THE STORY OF YELLOWSTONE

The year was 1806. Having reached the Pacific Ocean, William Lewis and Meriwether Clark now traveled homeward on the final leg of their famous journey. Somewhere in modern-day Montana, an expedition member named John Colter received permission to strike out on his own. A year later, he reappeared with tales of boiling lakes, bubbling bogs, and rivers flowing straight up to the sky. Colter's account of the Yellowstone region—the first by a Euro-American—was dismissed as nonsensical. Skeptics referred to Yellowstone as "Colter's Hell," a fictional land of fire and brimstone. Over the next fifty years this perception remained, nurtured by fantastical tales from explorers and adventurers headed for the Montana gold mines. Mountain man Jim Bridger, who had traveled through the region since the 1830s, described lakes where he could both catch and cook his dinner with the simple swing of a fishing rod from one pool to another.[4]

In 1870, to settle the issue, the surveyor general of the Montana Territory, Henry Washburn, lawyer Cornelius Hedges, and Montana politician Nathaniel Langford organized an expedition to the region. Provided with a military escort led by U.S. Army lieutenant Gustavus Doane, the Washburn Expedition followed upon the heels of an earlier expedition led by Charles W. Cook, David Folsom, and William Peterson in 1869. While the Cook Expedition provided early verifications of Yellowstone Lake, the thermal pools of the Lower Geyser Basin, and the Grand Canyon of the Yellowstone, the Washburn Expedition not only reaffirmed these sites but also provided the first "reputable" report on the presence of Yellowstone's most famous geysers, including one they christened "Old Faithful" (figures 4.2 and 4.3). Significantly, upon their return, Langford embarked on a speaking tour through eastern cities at which he confirmed many of Colter's original claims. A federal expedition soon followed, led by Ferdinand Hayden, head of the Geological and Geographical Survey of the U.S. Territories. In addition to scientists, Hayden invited painter Thomas Moran and photographer William Henry Jack-

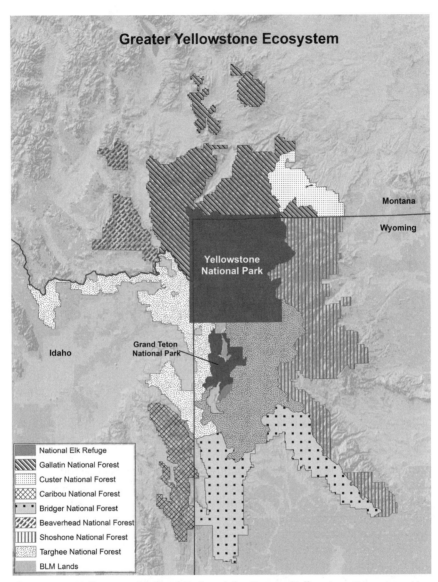

Greater Yellowstone Ecosystem

Montana

Wyoming

Yellowstone
National Park

Grand Teton
National Park

Idaho

National Elk Refuge
Gallatin National Forest
Custer National Forest
Caribou National Forest
Bridger National Forest
Beaverhead National Forest
Shoshone National Forest
Targhee National Forest
BLM Lands

Figure 4.2. Yellowstone National Park and the Greater Yellowstone Ecosystem. Map by Jessica Lee.

Figure 4.3. Old Faithful geyser in Yellowstone National Park. Photo by author.

son to record the journey. The images they brought back offered many Americans a first glimpse of Yellowstone's majestic landscapes.

In his final report, under strong urging by officials from the Northern Pacific Railroad, Hayden recommended the land be set aside as a public park. He noted that the high elevation of the Yellowstone region made it unsuitable for farming and ranching. The unstable landscape, caused by continuous geologic activity, rendered mining dangerous and likely unfruitful. Hayden went on to caution that failure to act quickly would allow Yellowstone to suffer the same fate as Niagara Falls. The distasteful commercial development around the falls had been heavily criticized by European dignitaries such as Alexis de Tocqueville, and it remained a point of embarrassment for the United States.

A bill soon appeared before Congress to retain Yellowstone under federal protection. Debate was minimal, but supporters reiterated the claim that the land held no commercial value and that setting it aside would cost the government nothing. Interestingly, the request for federal retention was not based on some grand vision of a national public land system but rather on the fact that Yellowstone lay within a U.S. Territory. Unlike Yosemite Valley, which was located within the bounds of an established state (California) and, therefore, was created as a state park in 1864, Wyoming was still a territory and, thus, national designation was required. With little fanfare, on March 1, 1872, President Grant signed the Yellowstone Park Act, creating the world's first national park.

It bears noting that Yellowstone did not represent the first federal action to set aside portions of the public domain for conservation purposes.[5] Nonetheless, its status as the first designated national park in the United States makes it unique in the history of conservation. Moreover, the story of Yellowstone's creation powerfully illustrates the influence of several conceptual tensions underlying national parks in the United States today. While the original impetus for protection may have been rooted in Romantic notions of inherent value in nature, the political debates—not surprisingly—played out on a much more pragmatic terrain. Congressional arguments focused on three central points, each of which addressed dominant conceptions of the day regarding nature and the public domain.

First, supporters highlighted the political role of natural wonders. As a source of national pride, the parks gave the young nation a response to the rich cultural or historical monuments found in Europe. The United States may not have prodigious museums or distinguished traditions in the arts, but it did contain natural monuments that rivaled or surpassed those found in Europe. Moreover, the move to protect Yellowstone suggested, however weakly, some measure of societal maturity and forethought that had been absent in the commercial defacing of Niagara Falls. In this way, advocates stressed the continuity between national sovereignty and the public domain: how the physical landscape could reflect not only the power but also the character of a modern nation-state.

Second, supporters argued that the preservation of Yellowstone would not cost the government one dime. This meant in practice that the park would have no budget and no professional managers. Instead, it would rely solely on volunteers. While this appealed to those concerned with government spending, this position also reflected adherence to the conception of public lands as unpeopled, static, and pristine. If the land was left untouched, the logic went, it would remain the same forever, as it had been since the colonial era. This proved problematic for a number of reasons. To begin with, it assumed that parklands would indeed go unmolested or modified by the simple act of designation. (However, as we shall see, this was never the case.) It also ignored the history of indigenous peoples in actively shaping precolonial landscapes into what many Euro-Americans assumed to be pristine wilderness. And finally, this conception failed to grasp a simple scientific truth—nature is intrinsically dynamic. While ecosystems are indeed relatively stable, nature writ large is nonetheless in a constant state of change. In short, without active management, the very landscape or wildlife characteristics for which a park was originally created could be at risk.

Third, and most importantly, park creation was premised on the notion that Yellowstone lacked economic value.[6] In this way, advocates addressed what was perhaps the greatest hurdle to park creation: the idea of nature-as-commodity. For over one hundred years, the primary purpose of the public domain was privatization and development, a viewpoint that remained no less prominent during the Yellowstone debates. All realized that preservation would likely have failed if the land were deemed suitable for settlement.

Interestingly, some of the most ardent park supporters—and those who argued most vehemently that Yellowstone lacked economic worth—were the railroad companies. The very same industry that benefitted most from federal policies designed to privatize the public domain now stood shoulder to shoulder with Romantic preservationists calling for park protection based on the land's inherent value. But clearly railroad officials believed that Yellowstone *did* have commercial value, just not in the conventional sense of farming, ranching, logging, or mining. Rather, they envisioned a relatively new market opportunity: nature tourism.[7]

The railroads already hauled agricultural goods and raw materials from the nation's interior to eastern cities. Park visitors represented a cargo for westbound trains. Upon arrival in Yellowstone, tourists could then serve as customers for railroad-owned hotels and restaurants. Though John Muir and other Romantics encouraged public visitation of national parks in order to build support in Congress, the railroads went a step further by actually facilitating the process, and in so doing, reaping economic rewards. This confluence of interest between the railroad industry and early preservationists helps explain the dual emphasis on resource protection and recreation that defines—and complicates—national park management today.

While these arguments were effective in convincing Congress to create Yellowstone, the conceptual contradictions on which they rested soon rendered effects of their own. Despite efforts to avoid the mistakes of Niagara Falls, Yellowstone quickly showed signs of suffering from the lack of proper management. The park director was a volunteer position first held by Nathaniel Langford. During his seven-year term, he visited the park only twice, setting a precedent for extremely lax oversight of Yellowstone's natural resources.[8] As a consequence, logging and hunting activity remained largely unregulated. Early settlers in Yellowstone—including the aptly named Yellowstone Improvement Company—set up monopoly control over the most scenic areas, charging admission and room and board to park visitors. Clearly, something needed to change if the new park was to endure.

One champion of such change was George Bird Grinnell, editor of the hunting and fishing magazine *Field and Stream*. Concerned with the rapid depletion of bison and other wildlife in Yellowstone, Grinnell teamed up with Senator George Vest of Missouri and General Philip Sheridan, big game hunter and commander of the U.S. Army in the American West. Together they sponsored a bill to strengthen regulations on hunting, to provide funds for park management, and to double the size of the already vast two-million-acre park to encompass the full migration routes of bison and elk. The bill failed, but they succeeded in appropriating funds for park management, reducing the land claims of the Yellowstone Improvement Company, and most importantly, passing an amendment allowing the Interior Secretary to call upon the U.S. military to enforce park regulations if needed.

In 1886, Congress decided once again to cut off all park funding. In response, the Interior Secretary contacted General Sheridan, who promptly dispatched the U.S. Cavalry to assume administrative control over Yellowstone. Though most assumed this arrangement to be short term, the army ended up serving as de facto park managers for the next thirty-two years!

JOHN MUIR AND YOSEMITE

For eighteen years after the creation of Yellowstone, political momentum for the establishment of additional national parks slowed considerably. While it's true that in 1875, Congress established the nation's second national park on Mackinaw Island in Michigan, within twenty years, the island was returned to Michigan and redesignated as a state park. Congress also authorized protection of Arizona's Casa Grande Ruins in 1889, but this site never achieved national park status.[9] Indeed, the establishment of new (and lasting) national parks would not occur until 1890, with the establishment of Yosemite, Sequoia, and General Grant National Parks in California. And these were achieved largely due to the efforts of one man: John Muir.[10]

Muir's family emigrated from Scotland to a Wisconsin farm when he was a young boy. After enduring a difficult childhood, Muir escaped to the University of Wisconsin where he studied geology and botany. He then gained employment in various manufacturing enterprises, discovering a skill for inventing mechanical devices and business management. But in 1867, on the verge of settling into an affluent career in industry, Muir abruptly changed his life course. An accident in a carriage factory left him temporarily blinded. Upon recovery, Muir resolved to set out on a one-thousand-mile trek to the Gulf of Mexico to experience God's creation. During the journey, he began to question the fundamental relationship between nature and human society. Rather than view nature as inherently inferior, valued only in terms of its human utility, Muir saw nature has having value in its own right.

Muir's insights echoed those of other American Romantics, such as Ralph Waldo Emerson and Henry David Thoreau, thinkers who also questioned the rationality of modern industrial progress. From a Romantic perspective, God's truest and most direct expression was reflected in unaltered wilderness, not in human-created cities or domesticated landscapes. Flipping the dominant viewpoint on its head, they found virtue and moral goodness rooted in inviolate nature. It was the human transformation of wild nature that was corrupt and unworthy; especially the urban industrial centers that served as hubs for greed, crime, and pollution.

Upon reaching Florida, Muir decided to sail to California to see a place he had read about called *Yo-Semite*. In 1869, after spending several months as a sheepherder in the Sierra Nevada high country, Muir arrived for the first time in Yosemite Valley. Five years earlier, in 1864, President Lincoln had signed a law granting the seven-square-mile Yosemite Valley to the state of California as a state park. The rationale for protection—that the land held no economic value—would echo strongly eight years later in the fight for Yellowstone. A group of state bureaucrats, collectively called the California Commission, administered the valley and appointed Galen Clark as the first "park guardian." But in practice, private entrepreneurs, such as James Hutchings, already controlled many of the most scenic areas, and set up unregulated businesses to capitalize on park visitation. As one of Yosemite's earliest promoters and largest tourism operators, Hutchings arrived in the valley in 1855. A few years later, he hired John Muir to run a sawmill in order to improve and expand his rustic inn. Muir took the job, but he also took every opportunity to wander and explore the mountains.

Word of Muir's exploits soon began to spread among visitors to the valley. In 1871, Ralph Waldo Emerson arrived and asked Muir for a tour. Two years later, at the urging of Emerson and others, Muir left the valley to begin writing articles about his adventures in Yosemite. After he published essays in *Harper's Magazine* and other outlets, Muir's fame, and that of Yosemite Valley, grew by leaps and bounds.

In 1879, Muir traveled north to visit Glacier Bay in Alaska. His published account of the journey introduced the bay to the American public. Soon, a tourism company brought visitors via steamship to see what was now called Muir Inlet and Muir Glacier. Upon his return to California, Muir married and began running his father-in-law's fruit orchard in Martinez, California. It would be ten years before he finally returned to Yosemite Valley. In 1889, a request for a tour by Robert Underwood Johnson, publisher of *Century Magazine*, provided the opportunity, but it did not prove to be a pleasant homecoming. Shocked by the scene of rampant commercialism

---

TEXTBOX 4.1.
John Muir: Mountain Poet

Imagine yourself in Hetch Hetchy on a sunny day in June, standing waist-deep in grass and flowers (as I have often stood), while the great pines sway dreamily with scarcely perceptible motion. Looking northward across the Valley you see a plain, grey granite cliff rising abruptly out of the gardens and groves to a height of 1,800 ft., and in front of it Tueeulala's silvery scarf burning with irised sun-fire. In the first white outburst at the head there is abundance of visible energy, but it is speedily hushed and concealed in divine repose and its tranquil progress to the base of the cliff is like that of a downy feather in a still room. Now observe the fineness and marvelous distinctness of the various sun-illuminated fabrics into which the water is woven; they sift and float from form to form down the face of the grand grey rock in so leisurely and unconfused a manner that you can examine their texture, and patterns and tones of color as you would a piece of embroidery held in the hand. Toward the top of the fall you see groups of booming, comet-like masses, their solid, white heads separate, their tails like combed silk interlacing among delicate grey and purple shadows, ever forming and dissolving, worn out by friction in their rush through the air. Most of these vanish a few hundred feet below the summit, changing to varied forms of cloud-like drapery. Near the bottom the width of the fall has increased from about twenty-five feet to a hundred feet. Here it is composed of yet finer tissues, and is still without a trace of disorder—air, water and sunlight woven into stuff that spirits might wear.

Source: John Muir, *The Yosemite* in *The Eight Wilderness Discovery Books* (Seattle: The Mountaineers, 1992), 712. First published in 1911.

---

taking place in the valley and the severe degradation caused by the overgrazing of sheep in the surrounding mountains ("hoofed locusts," as Muir called them), the two men decided to launch a campaign for a new national park. Focused on the area surrounding Yosemite Valley, they pitched the idea of a new park to politicians, railroad executives, and the general public. Approximately one year later, in 1890, President Benjamin Harrison signed a law creating three new national parks in California: Yosemite, Sequoia, and General Grant (later named Kings Canyon). Significantly, the new Yosemite National Park did not include Yosemite Valley, which was still owned by California. However, some 1,500 square miles surrounding the valley gained new federal protection.

Boosted by this success, Muir formed the Sierra Club in 1892, a hiking and advocacy organization designed to lobby for additional national parks, including the return of Yosemite Valley to Yosemite National Park. The club's first campaign, however, was to support a new national park around Mount Rainier. For this effort, club members joined forces with the National Geographic Society, the Northern Pacific Railroad, and local business leaders in Seattle. Employing the same logic used to justify both Yellowstone and Yosemite, that the land held no commercial value and would require no government spending, a law establishing Mount Rainier National Park was passed in March 1899.

TEDDY ROOSEVELT AND THE ANTIQUITIES ACT

At the turn of the century, the United States boasted five national parks (see table 4.1), but their future was far from certain. There was still no reliable budget and no federal agency specifically charged with managing them. On the twentieth anniversary of Yellowstone's creation in 1892, a bill was introduced in Congress to dismantle Yellowstone by reopening it for settlement.[11] The bill failed but underscored the fragility of park status.

Table 4.1. The First National Parks

National Park	Date Established	State(s)
Yellowstone	1871	WY, MT, ID
Yosemite[a]	1890	CA
Sequoia	1890	CA
Kings Canyon[b]	1890	CA
Mount Rainier	1899	WA
Crater Lake	1902	OR
Wind Cave	1903	SD
Mesa Verde	1906	CO
Glacier	1910	MT
Rocky Mountain	1915	CO
Hawaii Volcanoes	1916	HI
Haleakala	1916	HI
Lassen Volcanic	1916	CA
Denali[c]	1917	AK
Acadia[d]	1919	ME
Grand Canyon	1919	AZ
Zion	1919	UT

Notes: This list does not include parks that were later removed from the national park system.

[a]Yosemite Valley and Mariposa Grove were established as a California State Park in 1864. In 1890, Yosemite National Park was created on lands surrounding and excluding these areas, which were later returned to the federal government and added to the park in 1906.
[b]Originally established as General Grant National Park until renamed and expanded in 1940.
[c]Originally established as Mount McKinley National Park until renamed and expanded in 1980.
[d]Originally established as Lafayette National Park until renamed in 1929.

What the parks *did* have in their favor was the U.S. Cavalry. While continuing to serve as manager of Yellowstone, the army soon expanded its duties, taking administrative control in Yosemite, Sequoia, and General Grant National Parks in 1891, and adding Mount Rainier to its list in 1899. The Buffalo Soldiers, African American cavalrymen celebrated for their actions in the nineteenth-century Indian wars, played a prominent role in these park assignments. Captain Charles Young, one of the first black men to graduate from West Point, became the first to serve as superintendent of a national park when he took command at Sequoia and General Grant National Parks. Still, the legal authority of the cavalry was limited. Without criminal charges for illegal logging, grazing, hunting, or forest fires, the most the military could do was issue warnings, confiscate firearms, and, in some cases, expel offenders from the park. One of the most effective methods was to escort illegal sheepherders to one side of the park while driving the herd to the opposite side. Such creative methods helped the land to recover, slowly but steadily.

In 1901, a new U.S. president took office, one who would show keen interest in parks and wildlife protection. Back in 1883, at the age of twenty-four, Theodore Roosevelt had taken his first hunting trip to Yellowstone. A year later, following the sudden death of his wife and mother, he returned again to the West to ranch cattle in the Dakota Territory. A few years later, he met George Bird Grinnell and, in 1887, they formed the Boone and Crockett Club to promote big game hunting and wildlife conservation. In the years to follow, Roosevelt supported efforts to create the National Zoo in Washington, establish the Adirondack Forest Reserve in New York, and pass the 1894 law protecting Yellowstone's wildlife (see chapter 6). In 1898, the national fame he earned with the Rough Riders in the Spanish-American War helped him rise to the governor's mansion in New York. In 1900, William McKinley chose Roosevelt as running mate for his successful presidential campaign. And one year later, with President McKinley's assassination, Roosevelt became the twenty-sixth president of the United States.

Once in office, Roosevelt lost no time in supporting the parks. In 1902, he signed legislation establishing Crater Lake National Park in Oregon, and in 1903, Wind Cave National Park in South Dakota. That same year he embarked on a whirlwind tour of the American West, including stops at Yellowstone and Yosemite National Parks. In Yellowstone, he delivered a speech extolling the democratic values inherent in the national park idea. In Yosemite, Roosevelt requested John Muir to serve as his guide (figure 4.4). During the tour, the two men along with several park rangers escaped from the formal entourage to camp out alone under the stars. Sharing thoughts and ideas over the campfire, Muir offered his theory on the glacial formation of Yosemite Valley and explained why it

Figure 4.4. President Teddy Roosevelt and John Muir overlooking Yosemite Valley in 1903. Photo courtesy of the Library of Congress.

should be rescued from the abusive management of the state of California. By the time they rejoined the group, Muir and Roosevelt had become fast friends. Three years later, in 1906, Yosemite Valley and the Mariposa Grove of large sequoia trees were officially transferred back to federal ownership for inclusion in Yosemite National Park.

The year 1906 marked two other key events in the evolution of the national park system. These included passage of the Antiquities Act and the creation of Mesa Verde as the nation's ninth national park. As the first park created to preserve objects of historic and cultural interest as opposed to "natural wonders," Mesa Verde represented new priorities for park protection. Meanwhile, the Antiquities Act provided the president with unprecedented powers that would forever alter the manner, politics, and scope of public land protection in the United States.

The intertwining story of these two events began in 1889, when local cowboys, Richard Wetherill and Charlie Mason, stumbled upon ancient Puebloan cliff ruins on the Ute Mountain Ute Reservation in southwestern

Colorado. Wetherill's subsequent efforts to excavate, display, and sell various artifacts from the site attracted the attention of antiquities collectors from across the globe. One of these was Gustaf Nordenskiold, an affluent amateur archeologist from Sweden. Working with Wetherill, he extracted over six hundred artifacts from the cliff dwellings. Intending to ship them back to Europe, Nordenskiold was arrested at the railroad depot in Durango and charged with looting treasures of national interest. But because no law existed protecting such artifacts, he was set free. In 1893, he published the first scientific article on the cliff dwellings. Meanwhile, the Wetherill family launched a tourism business to Mesa Verde, as the oldest brother, Richard, excavated other sites in Colorado, Utah, Arizona, and New Mexico, including the famous Chaco Canyon. The Four Corners region (so named for the intersection of four state boundaries) was soon inundated with artifact hunters in search of items to collect and sell.

Concern over the rampant looting of the ruins spurred local residents to seek park protection for Mesa Verde. One of the leading advocates was Virginia McClurg, a well-known writer, poet, and activist for women's rights. In 1900 she organized the Colorado Cliff Dwellings Association (CDA), an offshoot of the Colorado State Federation of Women, to support protection of the cliff dwellings and other sites. However, as the campaign progressed, McClurg shifted her vision for Mesa Verde from a national park to a "women's park," exclusively controlled and managed by her organization. This change led to a split within the CDA. Vice President Lucy Peabody, along with many influential members, broke away and threw their full support to the national park campaign, which eventually prevailed.

While national park status helped secure the ruins in Mesa Verde,[12] other historic sites across the Southwest remained vulnerable to artifact hunters. Worried that the legislative process required for park creation was too slow to halt the looting, Representative John Lacy of Iowa proposed a new law. The Antiquities Act made it a federal crime to disturb sites of historical, cultural, and scientific significance to the United States. However, the true potency of this law derived from two additional features. First, it allowed the president to protect sites by declaring them national monuments via executive order. By sidestepping Congress, the president could act quickly to protect fragile areas. But it also meant that local and congressional politics no longer dominated decisions to set aside public lands.

A second key feature was the precise wording contained in the Act, which, due to several last-minute amendments, resulted in expanding exponentially the power and scope of the original law. One change removed the 640-acre limitation on the size of monuments to read instead "the smallest area compatible with proper care and management." This

TEXTBOX 4.2.
The American Antiquities Act of 1906

Sec. 2. [T]he President of the United States is hereby authorized, in his discretion, to declare by public proclamation historic landmarks, historic and
prehistoric structures, and other objects of historic or scientific interest that are
situated upon the lands owned or controlled by the Government of the United
States to be national monuments, and may reserve as a part thereof parcels
of land, the limits of which in all cases shall be confined to the smallest area
compatible with proper care and management of the objects to be protected.

has since been interpreted to allow the protection of sites totaling tens of
millions of acres. Another amendment added the word *scientific* to the
types of sites covered in the law. Though clearly intended to protect sites
containing objects of cultural and historic interest (e.g., antiquities), expanding this list to include objects of "scientific interest" opened the door
to protect lands based purely on their physical features (see textbox 4.2).

President Roosevelt immediately put these interpretations to the test in
September 1906, by using the Antiquities Act to create the very first national monument at Devil's Tower, Wyoming. In so doing, he chose to
preserve a natural wonder with scientific value, rather than focus on a site
of cultural or historical interest as most expected. In fact, of the eighteen
monuments declared during his presidency, only six directly protected
antiquities, including El Morro, Chaco Canyon, and the Gila Cliff Dwellings in New Mexico, and Tonto, Montezuma Castle, and Kino missions in
Arizona. While national monuments generally provide less protection
than a national park, they often serve as a stepping-stone to park status.
Six of Roosevelt's monuments would go on to become national parks,
including one that would stretch the interpretation of the Antiquities Act
to its very limit: the Grand Canyon.[13]

During his 1903 visit, President Roosevelt stated that the canyon should
be left untouched. Local politicians and businessmen, however, strongly
disagreed. Seeing great potential for mining and private enterprise, they
fervently opposed any idea of creating a new national park. In 1906, Roosevelt declared portions of the canyon as a federal game reserve. Then, in
1908, realizing congressional action unlikely, Roosevelt took the bold step
of declaring over eight hundred thousand acres (or 1,279 square miles) of
national forestland surrounding the Grand Canyon as a national monument, to be administered by the USDA Forest Service. He rationalized
that the canyon represented "an object of unusual scientific interest, being
the greatest eroded canyon within the United States." Again, without in-

clusion of the word *scientific* in the Antiquities Act, this action would have been impossible. Moreover, the declaration severely tested the "smallest area" clause. Determined to undo Roosevelt's order, critics kicked off a fight that ended with their defeat in the U.S. Supreme Court and the eventual creation of Grand Canyon National Park in 1919. On the eve of leaving office, on March 3, 1909, Roosevelt again evoked the Antiquities Act to declare another huge national monument for some six hundred thousand acres surrounding Mount Olympus in the state of Washington. Without question, President Roosevelt was one of the greatest supporters of the national parks, although for John Muir, there was one striking blemish on his friend's otherwise glittering record.

THE FIGHT FOR HETCH HETCHY

Since the 1890s, a plan had been considered to dam the Tuolumne River in order to augment the water and electricity supply for the city of San Francisco. In the aftermath of the famous 1906 earthquake and fire that destroyed much of the city, efforts to construct the dam received a groundswell of new public support. The problem was that the proposed dam site required the flooding of a place called Hetch Hetchy, a mountain valley lying within the boundaries of Yosemite National Park. John Muir considered Hetch Hetchy to be second only to Yosemite Valley in terms of its sublime qualities. But even more worrisome was the precedent it would set in terms of national park protection. If Congress allowed such massive and irreparable development to occur in one national park, what hope was there that any of the other national parks would be safe in the future?

Because the land was federally owned, the dam required congressional approval. Congress had granted this on several occasions, but each time Ethan Hitchcock, the Interior Secretary, blocked the project. By 1908, however, the political winds had changed. California's congressional delegation favored the dam, as did President Roosevelt's close friend and confidant, Gifford Pinchot. As an advocate of progressive conservation, Pinchot believed that public lands should be used for rational economic development, to provide the greatest good for the greatest number of people. And for Pinchot, the dam project clearly met these criteria.

On the other side, John Muir argued that the greatest good would be served by preserving the valley for future generations. For Muir, the populist appeals made by corporate enterprises were thinly veiled attempts to benefit themselves while ostensibly providing resources for the people. Some of Muir's most famous writing derives from his impassioned pleas to save Hetch Hetchy.

These temple destroyers, devotees of ravaging commercialism, seem to have a perfect contempt for Nature, and, instead of lifting their eyes to the God of the mountains, lift them to the Almighty Dollar.

Dam Hetchy Hetchy, as well dam for water tanks the people's cathedrals and churches, for no holier temple has ever been consecrated by the heart of man.[14]

Despite his best efforts, Muir was unsuccessful in persuading President Roosevelt, and the dam was approved. Before it was enacted, however, a new president came to office. Like Roosevelt, President William Taft traveled west to visit Yosemite. And like Roosevelt, he asked Muir to serve as his personal guide. Once again, Muir convinced a U.S. president that Yosemite should be protected, and Taft withdrew federal approval for the dam. In 1910, Taft again proved his support for parks by signing the law creating Glacier National Park in Montana.

However, in 1913, the tide changed yet again. Upon taking office, President Woodrow Wilson's choice for Interior Secretary was Franklin Lane, a former city attorney of San Francisco and an ardent supporter of the Hetch Hetchy dam. By December, a new bill cleared Congress and was quickly signed into law. One year later, Muir died of pneumonia, and some would say, a broken heart.

Significantly, the defeat at Hetch Hetchy did not lead to a rash of dam building in other national parks. As with Muir's death, the loss in Yosemite appeared to galvanize public support for park protection. Muir's legacy was noted in 1915, when President Wilson expanded the system by signing a bill creating Rocky Mountain National Park in Colorado. The leader of the campaign, Enos Mills, cited inspiration from Muir to do for the Rockies what Muir had done for the Sierras. Still, the fight for Hetch Hetchy demonstrated beyond a doubt the need for a federal agency whose sole purpose was the management and care of the parks. And in a move steeped in irony, Franklin Lane's appointment as Interior Secretary provided the first step in filling that need.

STEPHEN MATHER AND THE NATIONAL PARK SERVICE

"20 Mule Team Borax." Coining this phrase helped Stephen Mather amass a fortune in the business world. Taking an early retirement, Mather enjoyed spending time hiking in the Sierra Mountains near his California home. However, one summer, after a visit to Yosemite National Park, Mather was so appalled with the accommodations that he dashed off a letter of complaint to the Secretary of the Interior, his old college friend Franklin Lane. Lane's reply changed the course of national park history:

"Dear Steve, If you don't like the way the national parks are being run, why don't you come down to Washington and run them yourself?" Mather took up Lane's offer on condition that he work on a volunteer basis and stay no longer than one year. Despite his high-powered industrial career, Mather was a member of the Sierra Club. He even met John Muir during a hike in 1912. An intense individual who often suffered nervous breakdowns, Mather found the restful solace of nature a welcome respite from the stresses of the business world. In Washington, he applied his tremendous drive and innate personal and organizational skills to achieve his first goal: the creation of a National Park Service (figures 4.5 and 4.6).

Mather began by building a powerful coalition of support, including Gilbert Grosvenor of the National Geographic Society and George Lorimer of the *Saturday Evening Post*.[15] He hired Robert Sterling Yard, former editor of *Century Magazine*, to write articles and news stories espousing the wonders of the parks. With his assistant, Horace Albright, Mather lobbied the railroads to publish the *National Parks Portfolio*, a polished

Figure 4.5.   Stephen T. Mather, first director of the National Park Service, touring Yellowstone National Park in 1923. Photo courtesy of the National Park Service Historic Photo Collection, Harpers Ferry Center.

Figure 4.6. The 1915 dedication ceremony of Rocky Mountain National Park in Colorado was attended by several of the major figures central to the early success of the national park system. From left to right, Stephen T. Mather, Robert Sterling Yard, Acting Superintendent Trowbridge, the first Park Service photographer, Herford T. Cowling, and Horace M. Albright, who would later succeed Mather as NPS Director. Photo courtesy of the National Park Service Historic Photo Collection, Harpers Ferry Center.

volume with photos and glowing descriptions of each park and monument. He gave copies to every member of Congress and worked with the General Federation of Women's Clubs for national distribution.

The fiercest opposition came from the U.S. Forest Service, arguing that a National Park Service would merely duplicate their existing conservation efforts. But Mather countered that the two agencies were distinct. For the Forest Service, the priority was conservation, which meant the rational development of resources for human use. For the new Park Service, the priority would be preservation, meaning the protection of resources for future generations. Mather effectively made his case, and one year later, in 1916, President Wilson signed the National Park Service Act into law.

The contradictory mandate contained in the Act—to both preserve resources and encourage human recreation—directly reflected the ideals and strategies adopted by Mather and his team. To protect the parks, Mather felt he needed a larger constituency. And the best way to do that was to expand the system and encourage visitors. Eyeing powerful al-

TEXTBOX 4.3.
The National Park Service Act of 1916

Section 1. Be it enacted by the Senate and House of Representatives of the United States of America in Congress assembled, that there is hereby created in the Department of the Interior a service to be called the National Park Service. . . . The service thus established shall promote and regulate the use of the Federal areas known as national parks, monuments, and reservations . . . which purpose is to conserve the scenery and the natural and historic objects and the wildlife therein and to provide for the enjoyment of the same in such manner and by such means as will leave them unimpaired for the enjoyment of future generations.

Thus reads the critical text comprising what many view as a contradictory mandate for the National Park Service. However, later on in Section 3 of the Act, the Secretary of the Interior is also instructed to "make and publish such rules and regulations as he may deem necessary or proper for the use and management of the parks, monuments, and reservations under the jurisdiction of the National Park Service." The Secretary is also given the ability to "sell or dispose of timber . . . and provide in his discretion for the destruction of such animals and of such plant life as may be detrimental to the use of any of said parks, monuments, or reservations . . . grant privileges, leases, and permits for the use of land for the accommodation of visitors . . . [and] grant the privilege to graze livestock."

Together, these clauses grant tremendous discretion to the Interior Secretary to allow activities in national parks that might otherwise be restricted. The most glaring example of the controversial effects rendered by these loopholes was the historic battle over Hetch Hetchy Valley in Yosemite National Park in California.

lies in the transportation and tourism industries, Mather continued his overtures to the railroads and stressed the new economic opportunities available in the parks. The idea that the parks needed to be used in order to be saved, and the belief that commodification could lead to preservation, were controversial views that would never be fully resolved. In later years, such notions would exacerbate tensions between the Park Service and environmental advocates. But for now, they guided Mather as he pursued his dream of expanding the young system.

With an agency now in place, the next goal was to upgrade and increase the number of parks. To these ends, Mather often spent his own fortune, paying for the ranger headquarters building in Glacier National Park and purchasing, repairing, and donating the privately owned Tioga Road through Yosemite. He was also extremely successful in convincing the railroads to invest in the parks by building branch lines and providing upscale lodgings for visitors. As a result, the Northern Pacific financed the massive Old Faithful Inn in Yellowstone, while the Great Northern and Southern Pacific made similar investments in Glacier and Yosemite, respectively. Meanwhile, Mather lent enthusiastic support to tourism programs, such as the Northern Pacific's "See America First" campaign. Printed on every Northern Pacific brochure, timetable, and press release, officials hoped the slogan would lure well-to-do eastern vacationers away from European destinations. Glacier National Park was christened "America's Switzerland" and even included accommodations designed as Swiss chalets and employees outfitted in traditional Swiss costumes.

Mather's efforts paid off. In 1916, the same year that the National Park Service Act was passed, two new national parks were created. Hawaii National Park (later split into Haleakala National Park on the island of Maui and Hawaii Volcanoes National Park on the Big Island) helped protect the fragile vegetation and landforms surrounding the active and dormant volcanoes on the island chain. A week later, Congress created Lassen Volcanic National Park to protect similar geologic features in California's Cascade Range. And the following year, a two-year campaign initiated by the Boone and Crockett Club under the leadership of Charles Sheldon successfully concluded with the establishment of McKinley National Park in Alaska (later rechristened Denali). By far the largest park to date, McKinley was designed to protect not only the highest peak in North America but also the rich and abundant wildlife in the region.

Mather's next goals were twofold: to expand the park system eastward in order to cultivate a truly national constituency, and to make them accessible to even more people by accommodating a popular new form of transportation: the automobile. The U.S. Army had generally banned "horseless carriages" from the parks during its tenure as manager. Mather changed this by actively encouraging automobile traffic, building scenic motorways, and promoting a park-to-park national road system. By 1919, over ninety-eight thousand park visitors arrived by car, outnumbering train passengers four to one.[16] Automobiles also gave people more access to areas within the parks. No longer dependent on the railroad companies' open-air coaches, motorists could go where they pleased, when they pleased, and camp anywhere in reach. In Rocky Mountain National Park, Mather campaigned for the building of Trail Ridge Road, which reached elevations over eleven thousand feet. In Glacier National Park, the Going-to-the-Sun Road took visitors to Logan's Pass, offering glimpses of mountain goats and bighorn sheep.

However, not everyone was enthralled with Mather's overtures to the automobile. Fearing that the balance between preservation and use was becoming untenable, longtime colleague Robert Sterling Yard, now president of the National Parks Association, wrote articles questioning this policy (see chapter 8). Undeterred, Mather turned his attention to expanding the park system eastward. The challenge here was that most land east of the Mississippi was already either privately owned or controlled by the states. Therefore, any new park creation would first require funds to purchase the land. This was a completely new problem, requiring a different approach and set of rationales.

Off the coast of Maine lay Mount Desert Island, so named in the seventeenth century by French explorer Samuel de Champlain. Lying on the southern boundary of French colonial Acadia, by the late nineteenth century, the island had come under American control. It now served as a popular site for lavish summer homes occupied by America's wealthy elite. One such island resident was Charles Eliot, president of Harvard University. Inspired by the final wishes of his recently deceased son, Eliot worked with another resident, George Dorr, to form a trust designed to purchase and protect the wildest parts of the island as a national park. In 1916, having secured over five thousand acres, Dorr offered the land to the federal government on behalf of the trust. President Wilson accepted the land, but he was unable to garner sufficient congressional support for a national park. He therefore decided to designate the land under the Antiquities Act as Sieur de Monts National Monument. Still not satisfied, Eliot enlisted the aid of another influential resident in the campaign for park status—John D. Rockefeller Jr. By donating another $3.5 million to the effort, Rockefeller quadrupled the size of the original parcel and created a network of carriage roads for visitors. Finally, in 1919, Congress passed a law establishing Lafayette National Park (later renamed Acadia). It was the first park in the eastern United States and the first created from the purchase of private land.

That same year, President Wilson created two additional national parks: the Grand Canyon in Arizona and Zion in Utah. Both were originally national monuments declared by Teddy Roosevelt more than a decade before. And in 1921, growth continued with the creation of Hot Springs National Park in Arkansas (originally set aside as a reserve by Congress in 1832). In 1923, President Warren Harding declared Bryce Canyon National Monument, which was redesignated shortly thereafter as Utah National Park in 1924 and renamed Bryce Canyon National Park in 1928.

With a park now established in the Northeast and continued park expansion out West, Mather turned his attention southward. In 1926, a national search committee with strong backing by the respective state delegations led Congress to authorize the creation of three new parks in the southern United States: the Mammoth Caves in Kentucky, the Shenandoah Mountains in Virginia, and the Great Smoky Mountains in North Carolina and Tennessee. The caveat was that the states, rather than the federal government, must raise the funds to purchase the needed lands.

In the Smokies, the $10 million fund-raising goal seemed insurmountable given the poverty of the region. Still, piecemeal donations trickled in from groups and individuals, ranging from schoolchildren to wealthy patrons. Opposition came from two sources, the most powerful and vocal of which was the timber industry. Owning over 85 percent of the land within the defined boundaries of the new park, logging companies argued that it would cost jobs and damage the frail regional economy. Knowing the state governments of North Carolina and Tennessee would evoke the power of eminent domain to force the sale of the property, timber interests promptly logged their lands before (and sometimes illegally after) making deals.

The other source of opposition derived from several thousand residents who still called the mountains home. Many of these were impoverished Cherokee Indians or white residents who worked in the local coal mines. Forced to sell their meager homes and move elsewhere, few of these economically marginalized households could find alternative places to live that they could afford. While some went willingly, others fought to the bitter end in legal battles made all the more difficult by the lack of basic literacy skills or political influence. Ultimately, most resisters lost their cases in court.

Despite these efforts, by 1927, only half of the needed funds had been raised. With the timber rapidly disappearing, the gap was closed—once again—by a donation from John D. Rockefeller Jr. However, by 1929, the impacts of the Great Depression created another shortfall, as donors could no longer fulfill earlier pledges. To make up this new deficit, President Franklin D. Roosevelt stepped in to provide the remaining $1.5 million needed to finalize the deal. On July 3, 1936, FDR gave a speech at the official dedication ceremony for Shenandoah National Park. Four years later, the president spoke again, this time at the dedication for Great Smoky Mountains National Park. The dedication of Mammoth Caves National Park followed in 1941. Because of the challenge of acquiring private property for park creation, rather than simply redesignating portions of the public domain, the time between congressional authorization for a national park and its final dedication could be lengthy. In the three cases described above, the process spanned over ten years!

Today, Great Smoky Mountains National Park is the most visited national park in the country.[17] With over 9.6 million visitors in 2012, it received more than twice the number of the second most popular park, the Grand Canyon. And Shenandoah's Blue Ridge Parkway is the most visited "unit" of all unit types in the national park system,[18] with over fifteen million recreation visits in 2012. By such measures, the southern parks have been a tremendous success.

But at what cost? The legacy of these three southern parks remains controversial. For the first time, the image of public lands as unpeopled or "pristine" was shattered by the historical mining and logging activity as well as the continued presence of local white and indigenous residents, some of whom did not wish to leave. Of course, the conception of pristine nature as a basis for park creation has always been a fallacy, dependent upon the negotiated or forced dislocation of indigenous peoples in order to declare them "empty" and uninhabited prior to settlement or park designation. But at the time of congressional authorization for Great Smoky and Shenandoah, people were still there. And though many were illiterate and poor, residents still found ways to express their anger at what many viewed as an abuse of their basic rights. On the other hand, to frame this issue as a "federal land grab" also misses the point, given the leading role played by state governments, local organizations, and local residents who so strongly lobbied for park creation.

In the end, most local occupants moved out, some returned to illegally squat in what remained of their former homes, and a few individuals received official permission to stay. Over time, the Park Service began to acknowledge the history, culture, and personal sacrifices made by residents of these mountain communities. Today, their stories are increasingly incorporated into the educational and interpretative mission of the parks. In the future, federal agencies would adopt new collaborative approaches to public land planning, management, and designation; approaches designed to better accommodate local communities—including indigenous peoples—in conservation initiatives. But this was still decades away.

THE JACKSON HOLE CONFLICT AND POSTWAR EXPANSION

In 1929, after fourteen years of service, Stephen Mather stepped down as park director. During his tenure, he had established the National Park Service, doubled the size of the system, made it truly national in scope, and greatly enhanced services and accommodations to make the parks accessible to more people than ever before. Mather was replaced by his former assistant, Horace Albright, who continued the push to expand the system while maintaining Mather's high standards for inclusion. In 1930 and 1931, Albright oversaw the creation of Carlsbad Caverns National Park in New Mexico and Isle Royale National Park in Michigan's Lake Superior. He then persuaded President Herbert Hoover to declare a series of eleven new national monuments. Some of these were later redesignated as national parks, including Badlands in South Dakota, Arches in Utah, Great Sand Dunes and the Black Canyon of the Gunnison in Colorado, Saguaro and the Grand Canyon (a section unprotected by previous declaration, but later absorbed by the national park of the same name) in Arizona, and Death Valley in California.

In 1932 and again in 1933, Albright successfully lobbied for the transfer of a number of national monuments from the authority of the Forest Service to the National Park Service. These included Bandelier in New Mexico and Mount Olympus in Washington State, along with twelve other sites. Working to broaden the scope of the agency, Albright convinced newly elected president Franklin D. Roosevelt in 1933 to place all military parks, battlefields, memorials, and historic monuments within the jurisdiction of the Park Service. In one fell swoop, places such as the Gettysburg Battlefield, the Statue of Liberty, Mount Rushmore, and the National Mall in Washington, D.C., all came under agency stewardship (figure 4.7). The parks increasingly served as repositories of the nation's history and identity, reflecting not only our relation to nature but to each other as well.

The New Deal period witnessed another phase of park expansion and diversification under the leadership of FDR's Interior Secretary, Harold

Figure 4.7.   Gettysburg National Military Park in Pennsylvania represents the expansion of the NPS mandate in the 1930s to act as steward of places deemed significant to American history and national identity. Photo by author.

Ickes, and new Park Service directors Arno Cammerer (1933–1940) and Newton Drury (1940–1950). As one of the most successful public employment initiatives of the New Deal, the Civilian Conservation Corps (CCC) put as many as three hundred thousand men to work per year in camps located in national and state parks across the country. They built roads and trails, fought fires, controlled flooding and erosion, managed wildlife, planted trees, and constructed campgrounds. These improvements allowed even more recreational activities in the parks. But the increase in roads and motorized tourism worried environmentalists to such an extent that in 1935, the Wilderness Society was formed. This new advocacy organization campaigned for the designation of "true" wilderness areas within the public land system—places devoid of roads and permanent structures, and protected from significant human modification (see chapter 8).

Nonetheless, Ickes continued to emphasize what he termed the "democratization of nature" by creating new land unit designations that encouraged recreation. In 1936, the Boulder Dam Recreation Area was established around the human-made reservoir created by the dam. A precursor of what would later become National Recreation Areas in the 1960s, the reservoir (later renamed Lake Mead) was administered by the National Park Service through an interagency agreement with the U.S. Bureau of Reclamation. The following year, in 1937, Cape Hatteras National Seashore in North Carolina became the nation's first federally protected coastal area. Concerned with the rate of private development along the nation's coasts, Secretary Ickes was determined to provide public access to American beaches for all citizens. He hired Ansel Adams to create what would become iconic black-and-white photographs of the national parks, furthering their popularity among the American public.

Meanwhile, the park system continued to expand with the creation of Big Bend National Park in Texas (1935), Olympic in Washington (1938), and Kings Canyon in California (expanded and renamed from General Grant in 1940). And although Congress authorized it in 1934, Everglades National Park finally received its dedication in 1947 after a thirteen-year campaign by the state of Florida to secure the necessary lands. Though requiring a compromise that allowed for continued oil and gas development within park boundaries, Everglades marked another turning point for park protection with its focus on preserving wildlife habitat rather than conventionally picturesque or sublime vistas.

While the establishment of each of these parks involved political struggle, few came close to the scale and intensity of conflict that took place in Wyoming over the Grand Tetons during this period. Back in the 1920s, one of Horace Albright's major goals was to enlarge Yellowstone National Park to include the Grand Tetons and the surrounding valley near Jackson Hole. In 1929, Congress agreed to use national forestland to create

Grand Teton National Park, but it refused to consider the valley since most of it was privately held. On Albright's request, John D. Rockefeller Jr. stepped in once again, devising a strategy to quietly purchase the land and donate it to the Park Service. However, in 1930, word got out about the plan, setting off political fireworks.

Interestingly, local opponents insisted that an expanded federal regulatory presence in the region would impinge upon private property ownership and free enterprise. However, while making this claim, they simultaneously argued that Rockefeller, who now owned the property in question, could not use the land as *he* saw fit (e.g., gift it to the American people). The logical contradictions notwithstanding, over the next decade, Wyoming's political delegation successfully blocked any congressional action to create a new national park. FDR finally responded by declaring Jackson Hole National Monument in 1943. The local backlash was intense, likening the president's action to the Japanese attack on Pearl Harbor and Hitler's annexation of Austria. Wyoming's politicians successfully passed a bill through Congress abolishing the national monument, which FDR promptly vetoed. The state then filed a lawsuit in federal court, but it was dismissed. Even after FDR passed away in 1945, the conflict raged on.

Ultimately, a coalition of conservation groups including the Boone and Crockett Club, Sierra Club, Wilderness Society, and Izaak Walton League worked for a compromise bill that would allow the monument to be added to Grand Teton National Park on several conditions. First, Teton County was to be reimbursed for lost property taxes. Second, seasonal hunts of the region's elk herd would continue. And third, existing grazing rights would be grandfathered in to the new park. But most profoundly, the deal included an additional provision stating that the Antiquities Act could never again be used in the state of Wyoming. Thus, in 1950, the state with the world's first national park (Yellowstone), national forest reserve (Yellowstone Timberland Reserve, see chapter 5), and national monument (Devil's Tower) also became the first to permanently ban the president's unique power to declare future monuments. It would be thirty years before a U.S. president would once again use the Antiquities Act on such a large scale.

Despite the struggle in Wyoming, visitation to national parks exploded in the postwar era, bringing new strains to the perilous balance between preservation and use. With the end of wartime fuel rations, America's newly prosperous middle class traveled to the parks in record numbers. In 1948, 29.8 million people visited the parks. Eight years later, the total number of visitors had more than doubled, to 61.6 million.[19] In 1954, five national parks counted over one million visitors per year.[20] The parks were simply not prepared for the onslaught. Understaffed, with roads and services in disrepair, a new management plan was desperately needed. In

1956, Park Director Conrad Wirth proposed a ten-year program titled Mission 66. Designed to coincide with the Park Service's fiftieth anniversary, the plan called for some $787 million in park improvements with an emphasis on roads, services, and the construction of visitor centers.

For many of the environmental groups that had long supported the Park Service, the focus on roads that provided even greater access to the more remote areas in the parks was a step too far. The parks were in danger of being loved to death. However, even as they openly criticized the Mission 66 program, these same groups did not hesitate to defend the parks against new pressures for development. In the early 1950s, a proposed dam project in Dinosaur National Monument in Colorado threatened a repeat of the Hetch Hetchy controversy, with the same end result. The Sierra Club's new director, David Brower, built a broad coalition of conservation, hunting, and public policy organizations to fight the dam, ranging from the Wilderness Society to the General Federation of Women's Clubs. This time, unlike the Hetch Hetchy campaign, their efforts paid off and the dam project was abandoned.[21]

STEWART UDALL, JIMMY CARTER, AND ALASKA

By 1960, growth in park visitation showed no signs of slowing, reaching some eighty million per year. Amid the social turbulence of the new decade, Interior Secretary Stewart Udall was the constant driving force for supporting, expanding, and diversifying the national parks. Serving in both the Kennedy and Johnson administrations, Udall strongly supported the ideas espoused in Rachel Carson's *Silent Spring* and the burgeoning modern environmental movement. In 1963, he authored his own best-selling work on the topic, *The Quiet Crisis*. During his tenure as secretary, Udall worked for the transformation of the Petrified Forest in Arizona from a national monument to a park in 1962, the creation of Canyonlands National Park in Utah in 1964, and Redwood and North Cascades National Parks in California and Washington State, respectively, in 1968. In addition, Udall oversaw the designation of six new national monuments, including Biscayne Bay in Florida.

However, the greatest pressure on the park system during this period derived from new federal programs designed to promote recreation. In 1962, a report from the congressionally authorized Outdoor Recreation Resources Review Commission (ORRRC) declared a "crisis in outdoor recreation" in the United States. In response, Udall created the Bureau of Outdoor Recreation to develop potential remedies. He also worked with the new park director, George Hartzog (1964 to 1972), and President Lyndon B. Johnson to promote the concept of "new conservation": making recreational opportunities, especially in urban areas, a major priority for the National Park Service. By "bringing parks to the people," Udall and President Johnson redefined conservation as a program designed to provide public health benefits to a greater number of citizens.

Their efforts led to new recreation-oriented designations within the park system, as well as new mechanisms for funding them. Building on precedents set by Harold Ickes in the 1930s, the Johnson administration established National Recreation Areas (NRAs) as new formal units within the park system. As places where visitors could recreate in ways not otherwise permitted in national parks, Udall oversaw the creation of nine such units, including the Golden Gate National Recreation Area in San Francisco and the Gateway NRA near New York City. He urged President Johnson to establish eight additional national lakeshores and seashores and to throw political support behind the Wild and Scenic Rivers Act and National Trails Act, both of which passed in 1968. In addition to recreation, Udall also campaigned for more traditional environmental preservation measures, including the 1964 Wilderness Act and the creation of fifty-six new national wildlife refuges. Finally, Udall expanded the historical preservation role of the Park Service through the addition of twenty new historic sites. To address the recurrent problem of funding these new units, Udall played a key role in passing the 1965 Land and Water Conservation Fund Act. This law provided monies for the acquisition and development of lands for federal, state, and local protection through royalties collected from offshore oil and gas development.

Like other forms of environmental protection, the dawn of the Environmental Decade in the 1970s brought renewed bipartisan support to the national park system. In 1971, President Richard Nixon oversaw the creation of Capitol Reef and Arches National Parks in Utah (both previously national monuments) and congressional authorization for Voyageurs National Park in Minnesota (formally established in 1975). The following year, in an effort to protect (albeit weakly) Everglades National Park from a planned airport development on its northern border that would seriously degrade the park's source of fresh water, President Gerald Ford signed the law establishing Big Cypress National Preserve. This designation still allowed for mining, hunting, grazing, agriculture, and a variety of other land uses; however, it signaled the intent to hold the land for future potential park status.

With the rise of the Carter administration, however, the park system found presidential support of a magnitude unseen since the days of Teddy Roosevelt. In the second half of his single term, President Carter signed into law four new national parks: Badlands in South Dakota (1978), Theodore Roosevelt in North Dakota (1978), Biscayne in Florida (1980), and Channel Islands in California (1980). However, it was in Alaska that Carter would leave his most profound and lasting public land legacy.

Recall that several areas in Alaska had been set aside for federal protection earlier in the century, including Sitka National Monument in 1910, McKinley National Park in 1917, and Katmai and Glacier Bay National Monuments in 1918 and 1925, respectively. These designations were largely a response to development pressures in the wake of the Yukon gold rush. However, over the next few decades, this pressure dissipated with the onset of the Great Depression and World War II. It was not until Alaska gained statehood that the next big change occurred.

As part of Alaska's unique statehood bill, Congress allowed the state twenty-five years to select an unprecedented 104.6 million acres from the public domain within its borders that it could keep as state land for perpetuity.[22] In addition, the new state was granted permission to keep some 90 percent of royalties generated from oil and mining leases on all remaining federal lands within state borders. This figure went well beyond the 25 percent given to all other states. In fact, both provisions were unprecedented in U.S. history.

Meanwhile, indigenous peoples in Alaska were growing concerned that state interests, focusing primarily on lands with oil and gas development potential, would begin selecting traditional Native lands as well. With the discovery of oil in Prudhoe Bay and the announced intention to build an eight-hundred-mile trans-Alaska oil pipeline to Valdez in Prince William Sound, oil companies and their political supporters were eager to put aside any indigenous land claims that might hamper the project. In fact, both sides urgently wanted the matter resolved, though they sought different paths to resolution. Ultimately, the matter was settled with the Alaska Native Claims Settlement Act of 1971 (ANCSA). The law required that indigenous tribes relinquish all land claims in the state of Alaska, with the exception of forty-four million acres. In return, they received a cash payment of $962 million, half of which would be paid from oil revenues. Rather than deal directly with tribal governments, the law organized indigenous peoples into a series of regional corporations, to which the cash payments would be made. In this way, oil companies could proceed with the development of the Prudhoe Bay oil field as well as construction of the Trans-Alaska Pipeline without fear of reprisals from indigenous land claims.

Most significantly for national parks, Section 17(d)(2) of ANCSA gave the Secretary of the Interior two years to identify up to eighty million acres to withdraw as national parks, forests, wildlife refuges, or other forms of protected lands. However, if Congress did not ratify the selected lands by 1978, all would revert back to the public domain and become available for commercial development. The first proposal came to Congress in 1973, but it was overshadowed by the Watergate proceedings. Ultimately, the task came down to the wire. With just seven months to go, in April 1978, Representative Morris K. Udall of Arizona (brother of former interior secretary Stewart Udall) sponsored a bill to protect some 110 million acres as various parks, monuments, refuges, and preserves. The new preserves category was added to appease opponents fighting for hunting and recreational opportunities not allowed in parks. However, the threat of a filibuster by Alaska's two senators, Mike Gravel and Ted Stevens, killed the bill in the final minute. In response, President Carter made good on a promise to protect these lands so they would not revert to "exploitable" status. With time running out, Carter evoked the Antiquities Act to set aside over fifty-six million acres of national monuments. Simultaneously, the president instructed Interior Secretary Cecil Andrus to establish forty million acres of new and expanded wildlife refuges. One year later, in a move of solidarity with the Carter administration, the UN Educational, Scientific and Cultural Organization (UNESCO) designated many of these monuments as World Heritage Sites.[23]

Confronted with these unilateral actions, opponents of public land protection in Alaska were outraged. But Congress finally relented. As one of the first orders of business in 1980, Congress passed the Alaska National Interest Lands Conservation Act (ANILCA), setting aside 104.3 million acres, including forty-seven million acres of national parks, monuments, and preserves. The name of McKinley National Park was changed to Denali, meaning "the High One." The park was expanded by 2.4 million acres, while the original area was designated wilderness. Other new parks included Glacier Bay, Gates of the Arctic, and Wrangell-St. Elias. At 8.3 million acres (13.2 million including the adjacent preserve), Wrangell-St. Elias is the largest single national park in the United States. Lying adjacent to Kluane National Park in Canada, the combined area constitutes the largest contiguous area of protected land in the world. Though ANILCA contained some problematic compromises that have continued to define long-running conflicts over the Tongass National Forest (chapter 5) and the Arctic National Wildlife Refuge (chapter 6), its passage also marked the single greatest amount of land set aside for protection in U.S. history.[24]

FROM DEREGULATION TO
COLLABORATION . . . AND BACK AGAIN

In the decades following ANILCA, it became clear that the era of large-scale park expansion had come to a close. The 1980s ushered in a period of deregulation for public lands, and park creation was clearly not on the agenda. Neither President Reagan nor President George H. W. Bush used the Antiquities Act to create new monuments.[25] At best, the focus was on the renovation of existing parks, and at worst, attempting to sell them off to the highest bidder. Funds generated by the Land and Water Conservation Fund went unused, not only for "external" park expansion (creating new parks) but also for "internal" expansion by buying up privately held inholdings. Between 1981 and 1993, only two new parks were established, Great Basin in 1986, created from a national monument, and American Samoa in 1988, located on U.S. territorial islands in the South Pacific.[26]

Meanwhile, the persistent challenges of overuse, external pollution from expanding urban centers, and inholding development continued apace. Debates over noncompetitive contracts for concessionaires played out in Congress,[27] but for the public, the major concern focused on wildfires in Yellowstone. In the summer of 1988, nearly eight hundred thousand acres or 36 percent of the park burned, including some major tourist destinations. Confusion over the causes and consequences of the fires resulted in heavy criticism of Park Service fire management policies, which allowed certain fire events to burn in order to serve their natural ecological function on the landscape. While in fact decades of fire suppression and severe drought combined to create conditions for the massive blazes that year, visitors and the national media focused on immediate effects, including blackened landscapes and the fact that the park actually was forced to close for the first time in its history. Why didn't the NPS put these fires out when it had the chance? Public understanding of ecology and ecosystem dynamics remained clouded by more prevalent perceptions of the national parks as aesthetically pleasing playgrounds. With the most famous of these playgrounds now closed and damaged, the popular media was quick to blame the "risky" and "untried" management practices of federal agencies. Over the next few years, these concerns only appeared to grow. With the Exxon *Valdez* oil spill off the Alaskan coast in 1989 and the escalating conflict over the northern spotted owl in the Pacific Northwest, questions mounted over the federal government's ability to provide sound oversight for environmental matters.

Nonetheless, by the early 1990s, it appeared that the tide might be turning. President Clinton came to office in 1993 amid great optimism for a renewed commitment to national park protection and expansion. But despite pro-environmental appointments to key federal positions, Clinton's legislative achievements were less than expected. Nonetheless, under the advocacy of Interior Secretary and former Arizona governor Bruce Babbitt, the Clinton administration oversaw passage of the 1994 California Desert Protection Act, which established Death Valley and Joshua Tree National Parks (both from national monuments), designated sixty-nine new wilderness areas, and created the 1.4-million-acre Mojave National Preserve.

President Clinton also signed laws transforming a number of other national monuments into national parks, including Saguaro (1994), Black Canyon of the Gunnison (1999), Great Sand Dunes (2000), and Cuyahoga Valley (2000). In 1996, the president declared the 1.9-million-acre Grand Staircase-Escalante monument in Utah, the largest ever in terms of land area in U.S. history and the first to be managed by the Bureau of Land Management (see chapter 7). In the final days of his term, Clinton went on to declare nineteen more national monuments, totaling some 5.6 million acres. These ranged from the one-million-acre Grand Canyon–Parashant in Arizona to the 2.3-acre President Lincoln and Soldiers Home in Washington, D.C. In addition to his successes in expanding the park system, President Clinton also made significant efforts to protect parks from the impacts of mining operations on public lands and to support ecosystem-based management initiatives, including the reintroduction of wolves to Yellowstone Park and the protection of roadless areas on the national forests.

The new millennium witnessed a return to a deregulatory emphasis for public lands generally, and a retreat for park growth in particular, as the George W. Bush administration came to power. Congress created two national parks from former national monuments: Congaree in South Carolina and Great Sand Dunes in southern Colorado. Otherwise, as in the 1980s, the administration focused on reforming or rolling back existing environmental regulations. For example, within weeks of taking office, President Bush reversed President Clinton's ban on the use of snowmobiles in Yellowstone. He also initiated reassessments of endangered species listings for grey wolves and other species, worked to open the public domain to increased fossil fuel development, and, though legally bound to defend them, he indirectly encouraged lawsuits challenging President Clinton's national monument declarations (which the Supreme Court declared legal in a 2003 decision). As for national parks, the Bush administration emphasized renovation rather than expansion or resolving inholding problems through private land purchases. To this end, the visitor's center at Gettysburg National Military Park was completely refurbished in 2008.

Ironically, in President Bush's second term, he broke with his Republican predecessors and evoked the Antiquities Act to declare the largest marine protected areas in history. In 2006 he declared the 140,000-square-mile Papahanaumokuakea Marine Reserve, covering ten islands and atolls in the Northwest Hawaiian Islands, including Midway. Then, in the final weeks of his presidency, President Bush declared three new marine national monuments: the 95,000-square-mile Mariana

Trench, the 13,400-square-mile Rose Atoll, and the 86,800-square-mile Pacific Remote Islands.

These actions mark a stark contrast to the deregulatory emphasis of the Bush administration regarding environmental matters for most of its eight years in office. In the context of marine protection, the administration's successful bid to release the U.S. military from adhering to the 1972 Marine Mammals Protection Act quickly leaps to mind as one of several seemingly contradictory policy positions.[28] Despite these questions, the monument declarations do reflect a general tendency by outgoing American politicians to leave a "public land legacy" in their wake. However, in this case, it is perhaps more significant to note that the vast majority of President Bush's marine monuments consist of submerged lands located on the ocean floor. They are places where very few Americans will ever visit, and the declarations contain loopholes allowing for future resource development if deemed necessary. The new monuments required no money for acquisition, little additional funding for management purposes, and did not pose a threat to existing commercial interests. In short, they fit the bill on multiple conceptual fronts. The monuments represent new environmental protections while staying true to several dominant nineteenth-century conceptualizations, including the idea of nature-as-commodity and as unpeopled, eternal, and pristine. Despite the future challenges the monuments may pose for achieving lasting and more rigorous environmental protection, many environmental organizations still applauded the president's actions.

With the ascension of the Obama administration in 2009, the political pendulum swung back once again in support of environmental regulatory protection. The signature piece of public land legislation in his first term was the 2009 Public Land Omnibus Act, the largest land-based protection law in twenty-five years. The law designated over two million acres of public lands as wilderness areas in nine states, established over one thousand miles of wild and scenic rivers, and created three new national parks, three new national conservation areas, four national trails, and ten new national heritage areas. During his first term, President Obama also worked with his Interior Secretary, Ken Salazar, to rescind mining leases near Arches National Park and to prohibit uranium mining near the Grand Canyon. In 2012, the president declared April 21 to 29 to be National Park Week. However, that same year, the administration also agreed with Congress's decision to remove the grey wolf from the endangered species list to the consternation of many environmental groups who had championed the wolf's return to Yellowstone years before.

In March 2013, the president declared five new national monuments, including the 236,000-acre Rio Grande del Norte in New Mexico, the 1,000-acre San Juan Islands in Washington State, the Harriet Tubman

Underground Railroad Monument in Maryland, the First State Monument in Delaware, and the Charles Young Buffalo Soldiers Monument in Ohio.[29] It remains to be seen what new protections or expansions lie in store for the national park system in the final years of President Obama's second term.

Arguably, global climate change presents the most significant challenge to national parks in the coming decades. Already, evidence is mounting of climate-induced stress upon various wildlife and vegetative species in parks across the nation. The end result may be scenarios in which the species or landforms for which parks were originally established no longer exist within park borders (if they exist at all). Native species adapted to unique environments, such as the American pika, desert tortoise, or Sonoran pronghorn, are particularly vulnerable.[30] While wildlife may migrate to more northern latitudes for climatic reasons, there is no guarantee that the habitats they find in these places will be sufficient for their long-term viability. Forests, shrubs, and other forms of vegetation are also capable of migration, albeit at a much slower pace. But the likely result will be a varied series of mismatches between species and habitat. Only "generalist" species, those that are most able to adapt to a wide range of environments (including what we often refer to as "weeds" or invasive species), may truly prosper.

One of the most dramatic examples of the effects of climate change can be seen in Glacier National Park in Montana. Here, the receding glaciers offer one of the most vivid illustrations of climatic impacts. Though the park once contained over 150 glaciers when it was established in 1910, it currently has only twenty-five glaciers of significant size. Park scientists estimate that by 2030, the remaining glaciers will completely disappear.[31] Though the park was not created to preserve the glaciers per se (the name refers to the work of past glaciers in sculpting the dramatic mountain peaks and ridges), their presence has come to be synonymous with the park itself. The melting glaciers underscore the dynamism of nature, the need for human action, and the fallacy of adopting a static vision premised on the notion of nature as eternal and pristine.

Aside from climate change, what other issues lie on the horizon for the national park system? Given recent trends, it is unlikely that a new round of large-scale expansion is in store regardless of who is in office. As in the past, new parks will likely derive from existing national monuments. However, much of the most desirable portions of the remaining public domain have already been granted park or monument status. President Bush's marine monuments are a case in point. To find huge expanses of "new" public land to set aside, one may very well be forced to look in some new places . . . like under the sea.

A similar "nonexpansionist" trend can be found in park visitation. According to official National Park Service reports, in 1980, there were approximately 220 million visitors to the national parks. In 1987, this number reached an all-time high of 287 million. However, since then, the numbers have leveled off, fluctuating up and down every few years but exhibiting no linear pattern despite continued population growth in the United States. In 1990, the parks welcomed only 255 million visitors nationwide. In 1992, the figure rose to 274 million before falling again in 1996 to 265 million. Between 2000 and 2012 the annual number has bounced up and down between 266 and 285 million visitors. In 2012, visitation numbers reached 282 million.[32]

The bigger issue here is not necessarily one of numbers. While some may be concerned that more people should be visiting the parks, others focus on the heightened impacts of visitation. Coupled with the stresses presented by climate change, the indirect effects of new energy development projects, and the growing impacts of urban water and air pollution, recreational activities are presenting new threats to fragile ecosystems within park boundaries.[33] This brings us back to a one-hundred-year-old question: How can federal managers more effectively strike the balance between recreation and environmental preservation?

CASES

The history of the national park system demonstrates a number of recurring themes. Perhaps the single most pressing challenge facing park officials is how to address the dual mandate of managing for both recreation and preservation. This fundamental issue, in turn, directly reflects long-held conceptual assumptions underlying the public land system as a whole. As we have seen, support for park recreation derived not only from those wishing to cultivate a political constituency for the parks but also from those hoping to profit from nature tourism. The idea of nature-as-commodity was frequently evoked in these early debates and remains central to many of the most vexing management problems today. The issue of how best to manage snowmobile use in Yellowstone National Park provides a quintessential example of these tensions. In a similar fashion, the conception of nature as unpeopled, eternal, and pristine continues to inform and challenge park management priorities, often shaping the expectations of park visitors in profound ways. The clashing of such expectations against desires to pursue land use goals premised on nature-as-commodity thinking is clearly illustrated in the collaborative attempt to manage the bison herd at Yellowstone National Park.

Snowmobiles in Yellowstone

The tensions on either side of the snowmobile debate in Yellowstone National Park are as old as the contradictory mandate contained in the park's founding legislation—to simultaneously promote recreation and preserve the natural environment. In more concrete terms, the issue can be viewed as a direct outgrowth of Stephen Mather's decision in 1915 to promote motorized tourism. That year, most visitors to Yellowstone still arrived by train. But by 1930, over 90 percent of visitors reached the park by automobile. In the early 1960s, nearly two million people arrived in the park by car; however, most came during the summer months. In contrast, Yellowstone remained relatively quiet during the winter months. This changed in 1963, when the first snowmobiles arrived in the park. Viewing the machines as a good way to promote winter visitation, Park Service managers offered their support. Over the next few years, park managers began to groom snow-covered roads to make them more accessible for snowmobile enthusiasts. By the mid-1970s, approximately thirty thousand snowmobiles entered the park each year, and by 1992, the number nearly tripled, to some eighty-four thousand machines.

In 1995, after receiving hundreds of letters of complaint from park visitors, as well as grievances from park employees citing headaches and nausea on heavy snowmobile use days, Park Service officials initiated a two-year study of snowmobile impacts. Findings confirmed fears of poor air quality in the park. During the winter, levels of carbon monoxide failed to meet federal standards, and on the busiest snowmobile days, they registered the worst in the nation, outstripping concentrations found in heavily polluted urban areas near Los Angeles.

In addition to human health impacts, concerns were also raised regarding the potential negative effects of snowmobile use upon wildlife. For Yellowstone's bison, groomed winter roads facilitated greater movement, increasing the likelihood that the animals might cross park boundaries. This was particularly harmful since bison were no longer protected once they left the park. They could be shot by local ranchers fearful that the bison might transmit brucellosis to their livestock, a disease that could cause miscarriages in cattle.

After waiting two years for the Park Service to issue a decision to limit snowmobile access, an environmental group filed suit in 1997.[34] In a settlement, the Park Service agreed to begin working on a new winter management plan that would consider closing off parts of the park to snowmobiles. Both sides in the debate quickly mobilized. Motorized recreational groups, including national organizations such as the Blue Ribbon Coalition, stressed that snowmobiles allowed them to exercise the freedom to experience national parks in the way they pleased. Limits on snowmobile use, they argued, amounted to nothing less than a restriction of their rights as American citizens to access public lands. Snowmobile advocates also pointed to the importance of supporting the local tourism businesses and claimed that "extremist" environmentalists exaggerated the effects to wildlife and air quality. In contrast, they contended that snowmobile riders were responsible, environmentally caring citizens who made significant contributions to the local economy during the winter months.

Environmental advocates and ecologists responded that the air pollution, noise, and heavy presence of human visitors during the winter months put the health of wildlife at risk. Groomed snowmobile trails allowed bison to wander beyond park boundaries, putting their lives in jeopardy. Moreover, the machines allowed visitors to penetrate deep into the park, spreading noise and air pollution and disturbing elk and other wildlife in ways that might impair their continued viability. Environmental groups also pointed to health concerns for park rangers and visitors exposed to idling snowmobiles for long periods of time, as well as the economic costs of grooming and maintaining roads by the Park Service.

In 2000, the Park Service completed Yellowstone's winter management plan. With support from the Clinton administration, the plan called for a ban on all snowmobiles in the park effective April 2001. However, by that date, the Bush administration had come to power. President Bush's appointment for Secretary of the Interior was Gale Norton, a well-known proponent of recreational and industrial uses of public lands and resources. In one of his first administrative actions, Bush issued an executive order placing a sixty-day moratorium on the snowmobile ban. Meanwhile, recreational vehicle advocates filed a lawsuit challenging the Park Service's decision. Rather than defend the original 2000 decision in court, the Park Service agreed to conduct a new supplemental study. The results of this report overturned the ban and recommended an *increase* in the number of vehicles allowed in the park each day to 720. However, the new policy included a requirement that rented machines have the "best available technology" in order to reduce both noise and air pollution.

Since then, both sides have continued to file lawsuits. For its part, the Obama administration has adopted a compromise approach to the problem. Attempting to placate those who benefit from the commodity value of Yellowstone via nature tourism (just like the automobile industry, and before that, the railroads), President Obama is also seeking to ensure protection for Yellowstone's environmental resources. Since 2009, Yellowstone has operated according to a temporary winter plan that allows seventy-eight "snow coaches" and 318 commercially guided snowmobiles into the park each day. Like the Bush administration, the Obama policy requires that all machines must use the "best available technology." In 2010, the park began planning for a new long-term winter plan and collected nearly sixty thousand public comments on the issue. However, as of 2013, the planning process was still dragging on. Just like the larger conceptual divide between human use and environmental protection, or between nature-as-commodity and nature as something with inherent value, no final resolution has yet been found.

Bison Management

In the late nineteenth century, American bison were on the brink of extinction. Numbering at one time in the tens of millions, only a few hundred still existed by the mid-1880s. Yellowstone National Park represented one of the very few places in North America where bison had continually existed in their native habitat since prehistory. However, by 1902, their numbers at Yellowstone had dropped to less than fifty animals. In an effort to restore the population, the Park Service imported twenty-one bison from private herds. Over the next few decades, the bison population fluctuated wildly, in part due to "herd reduction" activities (e.g., annual hunts), which were finally halted in the mid-1960s.[35] By 1996, the population appeared to stabilize at around 3,200 head, a number most ecologists felt aligned with the size of habitat available at Yellowstone.

In the late 1990s, a collaborative team of officials from the National Park Service, Forest Service, USDA Animal Health Inspection Service, and the state of Montana began a planning process for the bison herds in the Greater Yellowstone Ecosystem area. The primary management problem centered on brucellosis, a disease that causes miscarriages in cattle and that can be toxic to humans as well. Ranchers worried that bison infected with brucellosis would transmit the disease to domestic cattle herds grazing outside the park boundaries. Historically, this has provided the justification for local ranchers to shoot and kill bison that wander across the line. The state of Montana became involved in the bison management plan in order to protect the interests of the state's beef producers, in particular the state's "brucellosis free" status.

In 2000, this collaborative group of federal and state officials issued the final environmental impact statement on the bison management problem, a document that detailed a full range of management options for Yellowstone. Alternatives included 1) allowing bison to roam freely over their historic range (both inside and outside the park), 2) managing bison numbers through controlled hunting seasons, 3) systematically capturing, testing, and slaughtering infected bison, 4) developing a brucellosis vaccine for bison, 5) purchasing additional winter range to reduce interaction with domestic cattle, and 6) quarantining infected bison in specific locations outside the park.

The chosen plan for management combined a number of these strategies. The overall goal was to maintain a bison population of approximately three thousand animals. Under the current plan, if any bison roam beyond park boundaries, the first step is to haze them back into the park or into designated holding areas. If hazing does not work, bison are captured and tested for brucellosis. If the total population is low (fewer than three thousand animals), the bison that test negative are released and those testing positive are slaughtered. If the total park population is high (over three thousand animals), then the responsible agency may choose to have them slaughtered regardless of the testing outcome. Concurrently, managers are implementing a vaccination program for all nearby cattle and all "eligible" bison (e.g., all calves and yearlings).

One of the most amazing aspects of this entire planning effort is that according to scientific evidence there has yet to be a single documented case of disease transmission from bison to domestic cattle under natural conditions. In other words, the risk of transmission based on historical records is extremely low. Moreover, scientists agree that the brucellosis present in the Yellowstone bison herd today *originated* from domestic cattle herds. That's right. It was the cattle that passed the disease to the bison in the first place. It would appear that the bison might have more to fear from domesticated livestock than vice versa.

Conceptually speaking, the argument that bison constitute a significant threat to the livestock industry powerfully underscores the dominance of nature-as-commodity thinking over other values of nature. For environmental activists, the degree to which federal agencies are willing to compromise and modify the management of several thousand bison in their historic range in order to protect beef producers from such a small risk of disease is nothing short of astonishing. Clearly, there are tremendous economic ramifications for Montana's beef producers if the state loses its brucellosis-free status. But given the scientific understanding of the risks involved and the focus on adapting bison to the needs of cattle rather than vice versa, the entire enterprise highlights the continuity of nineteenth-century conceptions of nature in the management of Yellowstone National Park. The impulse to preserve American bison may derive in part from a commodity-driven tourism motive as well. A highlight for many park visitors is the opportunity to see bison in the wild. But most environmental organizations argue that we should protect the Yellowstone bison on the basis of preserving biodiversity, the need to safeguard the cultural and natural heritage of the region, and because the creatures hold inherent value. Nonetheless, federal support for these values pales in comparison to the lengths managers are willing to go to accommodate commercial interests.

In a recent twist to this debate, new agreements have been reached to send some of the bison that wander off the park to new locations. In 2012, five hundred head were transported to a fenced pasture on Fort Peck Indian Reservation in Montana. This was the first time Yellowstone bison permanently left the park and the first time since the 1870s that wild bison have lived on reservation land.[36] Another eighty-seven head of Yellowstone bison caught outside of the park boundary were sent to Ted Turner's ranch in northern New Mexico. At his Flying D Ranch, Turner runs a herd of over fifty-five thousand bison for commercial beef production. According to his deal with the Park Service, the Yellowstone bison are to stay for five years, and Turner is allowed to keep any offspring they produce during that period to enhance the genetic diversity of his private herd.[37]

However, this arrangement has raised a number of additional questions from environmentalists. By transferring the bison out of Yellowstone, their status, and hence, their value, has changed. For at least some of these bison, they have gone from wild animals living under protected status to commodities valued as little more than "burger on the hoof" at Ted's Montana Grill restaurants.[38] The discordance between ecological and political boundaries that kicked off the whole debate—that the static, human-drawn boundaries of Yellowstone do not align with the dynamic, ecological geography of wild bison—is now joined with the constant blurring of conceptual boundaries, whereby deciding which bison are to be preserved and which consumed is a matter of where we decide to let them graze.

5

National Forests

No public forest reservation shall be established, except to improve
and protect the forest within the reservation, or for the purpose of
securing favorable conditions of water flows, and to furnish a continuous supply of timber for the use and necessities of citizens of the
United States.

—Forest Organic Act of 1897

[W]here conflicting interests must be reconciled, the question shall always be decided from the standpoint of the greatest good of the greatest
number in the long run.

—Gifford Pinchot[1]

Welcome to the (fill in blank) National Forest, Land of Many Uses.
If you've ever visited a national forest, you've seen them, standing as sentinels on the forest border. The ubiquitous brown, rhombus-
shaped signs serve as both an invitation and a reminder of the complex,
multifaceted role of this unique form of public lands. Although most
visitors may not give it much thought, that short secondary phrase marks
a world of difference between national forests and national parks, and
indeed, many other types of public lands (figure 5.1).

However, at times it can be hard to see this distinction. Like national
parks, national forests frequently serve as a recreational resource—a
place for camping, hiking, mountain biking, and fishing. Visitors hope
to experience the outdoors, catch a glimpse of wildlife, or perhaps commune with nature. And when encountered, forest rangers bring to mind

Figure 5.1.   The ubiquitous "Land of Many Uses" message, shown here posted on the boundary of the San Juan National Forest in Colorado (next to the author's daughter), underscores the multiple-use approach to resource management used by USDA Forest Service. Photo by author.

"Smokey Bear," a friendly and stalwart steward of nature, standing guard to protect not only the trees but also Bambi, Thumper, and the gang against the threat of wildfire. All of this sounds quite parklike.

Indeed, the millions of acres of breathtaking mountain ranges; thunderous rivers; bucolic streams; dense, green forests; flowering alpine meadows; and expansive arid valleys certainly exude a sense of wild nature preserved. In fact, some of the wildest places in the United States, if measured in terms of their official Wilderness Area designation, can be found within the boundaries of the national forest system or in areas that originated as such.[2] When you survey the changing aspen colors and wondrous, snowy mountain peaks of the Lizard Head Wilderness near Telluride, Colorado, you are gazing upon lands administered by the Forest Service. The same holds true for the million-acre Bob Marshall Wilderness in Montana, the Pemigewasset Wilderness in New Hampshire, or the Misty Fords in Alaska—all are situated on national forestlands.

In land area, the 193 million acres of the national forest system outstrip the eighty-four million acres of the national park system by more than two to one. And if one excludes Alaska, which accounts for a full two-thirds of national park acreage, the difference is even more striking. Within the lower forty-eight states, the national forest system is almost six times larger than all the national parks, monuments, and recreation areas combined. It is therefore likely that many American citizens may have visited a national forest, even if the names of these forests lack the iconic quality and promotional campaigns afforded to the national parks. Though visitor numbers to national forests are not tracked with the same precision that they are in national parks (forests do not have controlled entrances and fees), it is estimated that annual visitation to national forests exceeded 204 million in 2004.[3] In short, visitors to national forests could be forgiven for confusing national forests with national parks.

So what's the big deal about the phrase *Lands of Many Uses* on those forest signs anyway? How does this make national forests different? Consider the following: Let's say we're camping in a national forest and decide to hike up to a high-elevation alpine lake. After a strenuous climb, we top out on an eleven-thousand-foot ridge into a beautiful meadow, complete with bubbling brook and a smooth-as-glass, crystal blue lake. We have found the quintessential wilderness scene, only to look up and find a group of grazing sheep staring back at us. And these aren't Rocky Mountain bighorns, but domesticated Farmer-John-petting-zoo-style sheep like you might find on an Iowa farmstead.

Or perhaps while meandering across a river valley of wildflowers under the shadow of fourteen-thousand-foot peaks, we find ourselves dodging around droppings from a Hereford cow and calf (and perhaps the animals themselves). In addition to stares, this time we may receive the occasional "moo" and possibly notice barbed wire fencing, metal cattle guards in the roadway, or even a cowhand on horseback gathering strays. Each of these encounters reflects the notion of multiple use. A national forest is both a place for hiking and livestock grazing.

But it doesn't end there. As we walk (jog, mountain bike, drive a car) along a national forest road, we may come bumper to bumper with an eighteen-wheeler rambling along loaded down with freshly cut logs. Or we may pass by a team of off-road motorcycles, a string of pack horses, or, depending on the season, a group of hunters in orange vests, locked, loaded, and clinging to the backs of ATVs in search of elusive big game trophies. Occasionally, we may stumble upon signs declaring "no trespassing" pinned to a bit of wire fencing, marking either a parcel of private property or a mining claim. Indeed, ski resorts (with one-hundred-year leases), uranium mines, natural gas derricks, and logging operations may exist alongside solitary hikers, groups of Girl Scouts, firewood gatherers, and individuals foraging for wild plants. Ranchers herding cattle, wildlife ecologists counting songbirds, and lumberjacks

felling trees all reflect the notion of "multiple use" stamped into the rationale for the national forest system.

Now, it is critical to point out that not all uses occur on the same acre of
land. National forest managers are at pains to accommodate all of these
diverse activities in ways that cause the least amount of conflict among
users, while at the same time maintaining the health and well-being of the
land in their care. At least, that is the goal. Though as one might expect,
the task is monumentally complex and leaves much room for debate (and
sometimes lawsuits) over the most appropriate degree, timing, and location of various activities on any given acre of national forest.

Thus, while the national forest system does share ideological roots with
the national parks—both provide a counter to the unfettered nineteenth-
century development of environmental resources through the federal
retention of public lands—the forests in many ways represent the other
side of the conservation coin. Whereas the parks constantly grapple with
the recreation-preservation dualism, national forests more fully embrace
utilitarianism. But in so doing, they are no less immune to the conceptual
legacy of viewing nature as a commodity, or as "unpeopled, eternal, and
pristine," than are the national parks.

Currently, there are 155 national forests and grasslands in the United
States. They cover over 193 million acres and lie in forty-two states and
Puerto Rico (figure 5.2). These lands account for approximately 20 percent
of forested areas in the United States. In other words, the vast majority
of forested land in the United States (some 80 percent) remains privately
owned or managed by state or municipal governments. Nonetheless, the
question of how these federal lands should be managed—whether for
timber production, mineral development, livestock grazing, recreational
activities, or ecological preservation—has placed the national forests at
the forefront of environmental debates in the United States for most of
their history. The agency responsible for these decisions, the U.S. Forest
Service, is the oldest of the federal land agencies discussed in this book.
Perhaps most importantly, the agency is located within the Department
of Agriculture, rather than the Department of the Interior like all other
federal land agencies.[4] This distinction explains why managers insist on
referring to their agency as the USDA Forest Service and underscores,
at least in part, the focus on resource development. Both the agency's
unique position within the federal government structure and its commitment to multiple-use scientific management derive from the Progressive
Era conservation roots of the national forest system. Time and again, the
conceptual tensions described above have played out as the Forest Service
grapples over what kind of resource development to allow in which location, to what degree, by what means, and for whose benefit.

National Forest System

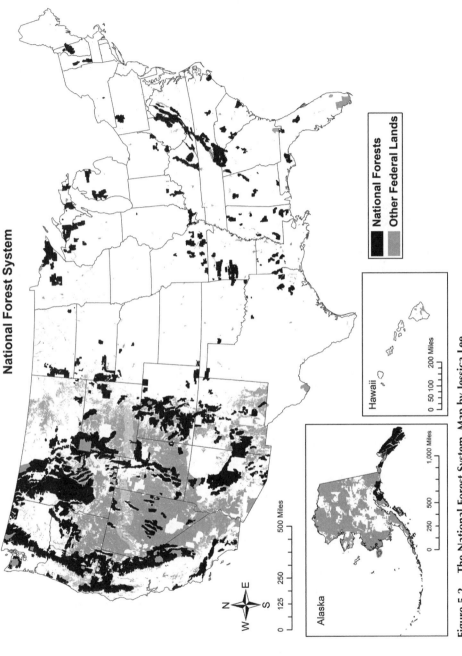

Figure 5.2.   The National Forest System. Map by Jessica Lee.

THE FIRST FOREST RESERVES

Scholars often trace the conceptual beginnings of progressive conservation back to the 1864 publication of George Perkins Marsh's seminal work, *Man and Nature, or Physical Geography as Modified by Human Action*. Serving as President Lincoln's ambassador to Italy, Marsh spent a great deal of time exploring reasons for the fall of the Roman Empire and other civilizations around the world. His book chronicled how deforestation, soil erosion, and general land degradation led to the collapse of these ancient societies. He went on to advocate for the careful, scientific management of resources to prevent a similar fate befalling the United States. As he noted, "Man has too long forgotten that the earth was given to him for usufruct alone, not for consumption, still less for profligate waste."[5] Marsh's work has been identified as one of the earliest articulations of ecological science in the United States, and it was highly influential in its own time among American natural scientists.

In 1873, former superintendent of the U.S. Census Franklin Hough presented a paper at the American Association for the Advancement of Science (AAAS) titled, "On the Duty of Government in the Preservation of Forests." Using Census data to demonstrate the deteriorating conditions of timbered lands in the United States, and drawing heavily upon Marsh's ideas, Hough laid out the rationale for the federal retention and management of forested public lands. Though it did not immediately impact policy, the paper brought attention to forestry issues and to Hough himself.

Two years later, in 1875, the American Forestry Association was established as an offshoot of the AAAS. The following year, Hough was appointed as the first federal forest agent. His subsequent reports on timber production and supply in the nation's forests fostered congressional support for the creation of the Division of Forestry within the Department of Agriculture. In 1881, the Division was established with Hough named as the first chief.[6]

However, for the first twenty-six years of its existence, the new agency had no actual land to manage or use for research. Congress viewed the division's role as purely advisory, providing information and guidance to private landowners upon request. Since the 1870s, efforts to set aside forested lands for actual federal protection had failed. Two Interior Secretaries, Carl Schurz, and later, William Sparks, both agitated to reform public land laws, which, in their view, were being exploited via fraudulent claims by loggers and land speculators. In addition, both men proposed the federal retention of some forestlands on the public domain during their respective tenures. But unlike the legislative success of Yellowstone National Park, their appeals fell on deaf ears in Congress.

Interestingly, many of those who did favor federal protection of the nation's forests adhered to a vision of preservation reflected in the national park ideal. In 1885, the State of New York set aside the Adirondack Forest Preserve, but with the goal of restricting all development, rather than the continuation of timber harvests via scientific management. For those wishing to provide a measure of protection for the nation's timberlands, it appeared as though there were only two options: appeal to private landowners or pursue federal intervention to stop all development. The utilitarian conception of conservation had yet to find a public audience or political traction.

Nonetheless, as newspapers chronicled the devastating effects of massive fires in the Rocky Mountains and Great Lakes region in the late 1800s, public awareness of problems facing the nation's forests began to build. The fires resulted in part from new logging technologies—saws and millworks that not only enabled but also required loggers to cut huge volumes of trees in order to keep them running at full capacity and profitability. Logging companies left massive slash heaps of branches, cuttings, and unwanted logs in their wake. Arid conditions quickly transformed these dry heaps into highly flammable tinderboxes, needing only a lightning spark or stray campfire to set them off. In 1871, the Peshtigo Fire in Wisconsin killed over 1,500 people, making it the deadliest in U.S. history.[7] Ten years later, a massive fire in Michigan killed some 160 residents. In both cases, the flood events that followed were even more destructive for local communities than the original fires.

Finally, in 1891, sufficient political support amassed to reform some of the most heavily abused western public land laws. As mentioned in chapter 3, the General Revision Act was designed to repeal the 1873 Timber Culture Act and to revise the 1876 Desert Lands Act. But in the final hours of debate on the bill, with the strong backing of George Bird Grinnell and new Division of Forestry chief Bernhard Fernow, Interior Secretary John Noble successfully lobbied congressional politicians to attach a rider. Though barely mentioned in congressional debates, the rider granted the president of the United States the unprecedented power to set aside forested lands in the public domain as "public reservations" via executive order.[8]

Within weeks of the bill's passage, President Harrison exercised his new powers to create the first forest reserve in the United States: the Yellowstone Timberland Reserve on the southern border of Yellowstone National Park (now part of the Shoshone and Bridger-Teton National Forests and Teton Wilderness). He went on to create the White River Plateau, Pikes Peak, South Platte, Plum Creek, and Battlement Mesa Forest Reserves in Colorado. By the end of his term in January 1893, President Harrison had declared fifteen forest reserves, covering more than thirteen million acres of land. While supporters generally believed that the forest reserves would help stem the tide of massive wildfires and corrupt public land practices, a great deal of confusion remained regarding their long-term purpose. Were they essentially a new form of national park, or something completely different?

THE 1897 FOREST ORGANIC ACT

Harrison's successor, President Grover Cleveland, followed suit by establishing an additional five million acres of reserves in Oregon in September 1893, including the enormous Cascade Range Forest Reserve and the much smaller Ashland Reserve. However, in response to growing congressional criticism, he refused to set aside any additional reserves until an explicit management policy was put into place. With Congress stalemated on the issue, Interior Secretary Hoke Smith and Wolcott Gibbs, head of the National Academy of Sciences, created the National Forest Commission to provide expert guidance. Dispatched out West in 1876, the Commission was charged with assessing the role of existing reserves, articulating management policy, and making recommendations on potential additional reserves.[9] Among the Commission's members were John Muir and thirty-year-old Gifford Pinchot, the first American-born, formally trained forester in the United States.

However, due to internal debates, the Commission initially failed to offer any coherent management plan. One point of dispute concerned who should manage the reserves: the U.S. Army (currently administering Yellowstone) or a corps of civilians trained in scientific forestry? With these matters unresolved, the Commission's first report focused on the one issue upon which all could agree: expanding the system. And President Cleveland happily obliged. In the final weeks of his presidency, on February 22, 1897, he declared thirteen new forest reserves, totaling more than twenty-one million acres. Known as the "Washington's Birthday Reserves," they more than doubled the size of the system and triggered an intense backlash by western politicians, ranchers, and loggers who assumed the lands would be permanently closed to development.[10] Congressional efforts to rescind the reserves through appropriation riders failed, including a pocket veto by President Cleveland on his last day in office.

Then in May 1897, the National Forest Commission submitted its final report. This time it included policy recommendations that emphasized the role of the forest reserves in protecting water supplies and underscored the fact that they should be used for timber production. This put to rest some of the worst fears held by western politicians. Shortly thereafter, a forest amendment was offered to an annual appropriations bill signed by President McKinley. Known as the Forest Organic Act, the amendment stated that American forest reserves were intended to secure "favorable conditions of water flows, and to furnish a continuous supply of timber for the use and necessities of citizens of the United States." It called upon the U.S. Geological Survey (USGS) to conduct surveys to assess existing and potential future reserves in the public domain. Lands deemed more valuable for mining or agriculture could not be included. To appease western lawmakers, the Act also included a repeal clause. With sixty days' notice and presidential approval, the Interior Secretary could recommend the return of existing forest reserve lands to the public domain if the land was found to be "better adapted" for other purposes. Additionally, all forest reserves remained open to future mineral prospecting and development.

Existing property owners were also protected. The law contained a "lieu lands clause," which allowed landowners who so wished to trade lands located within a forest reserve for lands of equal value located elsewhere. Those choosing to keep their holdings retained right of access to their property. This arrangement became a windfall for timber companies, who often chose to cut over their lands before trading them to the federal government for new forested lands located elsewhere. According to Zaslowsky and Watkins, the Weyerhaeuser company used the lieu lands clause to gain over nine hundred thousand acres of untouched forest lands.[11] Finally, the Act stated that all water resources, all timber, and all stone found on the reserves could be used "free of charge by bona fide settlers, miners, residents, and prospectors for minerals, for firewood, fencing, buildings, mining, prospecting, and other domestic purposes." The utilitarian priorities were clear: the forest reserves were to be used, not preserved.

However, they were also subject to government oversight. The Act identified the Department of the Interior as the "active managing authority," charged with carrying out its duties according to systematic, objective, and scientifically driven principles. The Interior Secretary was not only required to protect the reserves from "destruction by fire" but "for the purpose of preserving the living and growing timber and promoting the younger growth on forest reservations." Moreover, it must develop a systematic process to designate, appraise, mark, and sell "dead, matured, or large growth trees." This was a clear nod to modern science-based forest management. In addition, those appointed to carry out these tasks could "not [be] interested in the purchase or removal of such timber, nor in the employment of the purchaser thereof." For some, including Gifford Pinchot, this statement implied not only the need for modern forest management but also its implementation by a government agency uncorrupted by private industry and supplied with sufficient technical knowledge and moral fortitude to serve the public good. The only problem was that no such agency yet existed in the Interior Department. The one office that did fit the above description to a T—according to Pinchot—was the Division of Forestry, located not in the Department of the Interior, but in the Department of Agriculture.

GIFFORD PINCHOT AND THE USDA FOREST SERVICE

In 1898, Bernhard Fernow stepped down as chief of the Forestry Division to become dean of Cornell's new College of Forestry: the first four-year forestry program in the nation. Gifford Pinchot replaced him as chief (figure 5.3). Though the United States now boasted some forty million acres of forest reserves, the Division of Forestry was still without a single acre of its own to manage or use for research. Its primary task continued to be acting as a consultant to private timber companies and landowners.

Figure 5.3.   Gifford Pinchot served as first chief of the USDA Forest Service. He was largely responsible for establishing and applying the principles of progressive conservation to modern forestry in the United States. Photo courtesy of the Library of Congress.

The reserves, meanwhile, were overseen by two agencies in the Interior Department. As directed by the Forest Organic Act, a new division of Geography and Forestry was created to conduct forest surveys for the U.S. Geological Survey. Meanwhile, the General Land Office (GLO), an agency originally established to manage the sale and transfer of the public domain into private hands, attempted to take on the management of the nation's forest reserves as well. But according to Pinchot, the GLO officers were ill suited to the task. To ensure the well-being of the reserves, therefore, it was vitally important to transfer managerial authority to the Division of Forestry within the Agriculture Department. And Pinchot was just the right man to see it done.

Born to a wealthy family, Pinchot was raised to be a man of influence. The family fortune derived in part from land speculation, including the purchase of forested lands, which were then logged and quickly sold for profit as agricultural farmsteads. Graduating from Yale, Pinchot traveled to Europe to study forestry. Upon his return, he established himself as a private forestry consultant, managing the Vanderbilt family's seven-thousand-acre Biltmore Estate. After participating in the National Forest Commission and serving as a confidential forest consultant to the Interior Secretary, Pinchot entered full-time government service when he agreed to become the new chief of the Agriculture Department's Forestry Division. In Washington, he skillfully parlayed his family ties, remarkable drive, and charismatic personality to achieve extraordinary political influence. His close friendship with Theodore Roosevelt, dating back to Teddy's days as New York governor, transformed into a powerful asset when Roosevelt assumed the presidency in 1901.¹²

That same year, an agreement was brokered between the Departments of Interior and Agriculture over management of the forest reserves. While Interior continued to wield final administrative authority, the Division of Forestry was allowed to assess the reserves, offering technical assistance and management recommendations. But for Pinchot, it was still not enough. After years of heavy lobbying and Roosevelt's successful 1904 election, a bill finally passed Congress transferring all sixty-three million acres of the nation's forest reserves to Pinchot's Division of Forestry. The Transfer Act of 1905 also allowed the division to retain all profits from sales of timber and other forest reserve products, along with any user fees collected. Shortly thereafter, the Division of Forestry was renamed the Forest Service to underscore the goal of serving the American people. In a similar fashion, the national forest reserves were rechristened as the "national forests" to stress the fact that they were to be used for the nation's benefit, rather than be "reserved" like a national park.

As first chief of the new Forest Service, Pinchot worked tirelessly to promote the concept of conservation, an idea he began developing since first joining government service in 1898. In his 1911 memoir, *The Fight for Conservation*, Pinchot described the three principles of conservation this way:

> The first principle of conservation is development, the use of the natural resources now existing on this continent for the benefit of the people who live here now. . . . In the second place conservation stands for prevention of waste . . . [and third] The natural resources must be developed and preserved for the benefit of the many, and not merely for the profit of a few.[13]

It is telling that Pinchot worked to transfer the forest reserves to the Department of Agriculture rather than move his office to the Interior. For Pinchot, forestry was best understood as a form of agriculture, in which producers farmed trees. Timber was a crop like any other and should be managed accordingly. This often meant that management plans should maximize the most economically desirable tree species at the expense of others. However, he also advocated multiple uses of the forests: for livestock grazing, mineral extraction, and other societal benefits.

Pinchot summed up these ideas in the famous phrase, "the greatest good of the greatest number for the longest time."[14] Significantly, the idea of conservation transcended natural resource management. For Pinchot, Roosevelt, and other Progressives, it was a nationalistic, patriotic, and democratic cause; a nation was only as strong and independent as its resource base. Thus, efforts to prolong and ensure the quality of national resource supplies in the United States were vitally important to the well-being of the nation. Likewise, managing natural resources for the common good was highly democratic, allowing the greatest number of citizens to benefit from the nation's bounty.[15]

The use of scientific rationality and government intervention to overcome the wastefulness and greed of unfettered corporate action resonated strongly with Roosevelt's own political agenda. Having witnessed firsthand the effects of uncontrolled resource exploitation during his years ranching and traveling in the American West, Roosevelt shared Pinchot's passion for rational resource development. This same logic aligned with Roosevelt's Progressive agenda to rein in corporate monopolies that threatened the nation's economic and social fabric. For Roosevelt, strong, effective, and rational government action was the answer to many of the problems facing the United States at the turn of the nineteenth century.

To put these ideas into practice, Pinchot carefully built the new Forest Service from the ground up. To staff the agency, he sought out young men scientifically trained and highly motivated to serve the public good. Pinchot worked hard to develop an esprit de corps unmatched in any other federal agency. Frequently inviting Forest Service officers to his Pennsylvania estate, Pinchot arranged for seminars with distinguished guests and opportunities for social interaction to build agency cohesion. He wanted his men to be above reproach. They should be hardy enough to live alone in rustic conditions out West, with the courage to stand up to potential transgressors, but at the same time, Pinchot wanted them equipped with the technical skills to conduct rigorous scientific forestry.

To help instill these qualities in the burgeoning profession, Pinchot convinced his family to donate funds to endow the School of Forestry at Yale University. In 1900, the school opened its doors, directed by Pinchot's close friend, Henry Graves. Shortly after, Pinchot helped establish the Society of American Foresters. Serving as the Society's first president, Pinchot worked to promote the science and profession of forestry nationwide. Meanwhile, Yale's forestry program grew in stature. Though not the first such school in the United States, Yale soon became one of the most influential programs in guiding and shaping national forest policy. Pinchot relied heavily upon the school to furnish the Forest Service with many of its new employees.

While Pinchot focused on building the federal agency, President Teddy Roosevelt applied his considerable political power and influence to the task of growing the national forest system. He is credited with creating or expanding approximately 150 national forests (many of which were later consolidated), thereby increasing the system by a total of eighty million acres. Though permits to graze livestock on national forestlands were first issued free to stock growers in 1898, the Roosevelt administration—under prodding from Pinchot—decided to levy fees to cover administrative costs in 1906.

Of course, Roosevelt was similarly active in a host of other public land initiatives: establishing national parks and monuments, setting aside federal coal reserves, and instigating massive reclamation projects. In response to these combined actions, western regional sentiment again began to mount against any additional federal retention of public domain lands. To appease some of these concerns, Roosevelt signed a law transferring 10 percent of the proceeds from timber sales on national forestlands to county governments in which they were located to offset lost property taxes. Two years later, Congress increased the amount given to local governments to 25 percent of timber receipts. This move demonstrated concern for local interests. But at the same time, it produced an incentive for local and state politicians to pressure the Forest Service to maximize timber harvests for short-term economic gains, thus setting the stage for potential future conflicts among stakeholders over forest management priorities.

In 1907, Congress moved once again to appease western political and economic interests by including an amendment in the Agriculture Department's appropriations bill that rescinded the presidential power to declare or expand any additional national forests in the Northwest (Washington, Oregon, Idaho, Montana, Wyoming, or Colorado—though it would remain legal in other states).[16] Politically, Roosevelt felt he could not afford to veto the bill. However, on the eve of signing it, Roosevelt declared thirty-two additional or expanded national forests covering more than sixteen million acres in the affected states. Dubbed the "Midnight Reserves," this action enraged western politicians, but there was nothing they could do. Lawsuits brought by western timber and livestock interests were tossed out by the Supreme Court, reaffirming the government's right to both declare and manage the national forests according to the 1891 Forest Reserve Act and 1897 Forest Organic Act.

A BURNING ISSUE: FIRE POLICY

As with grazing fees and timber receipts paid to local governments, questions over fire policy presented another management challenge to the Forest Service that would render long-term conceptual and political tensions. The notion that fire presented a threat to forest environments was one of the primary rationales for the creation of the national forest reserves in the first place. But this perception would gain even greater influence as the Forest Service adopted fire suppression as one of its major management goals in the decades following Pinchot's departure from the agency.

After Teddy Roosevelt left office in 1909, his successor, President Taft, kept Pinchot on as chief forester for one more year before firing him for insubordination.[17] Though he had done much to develop, expand, and shape the national forest system and its governing agency, Pinchot's tremendous influence during the Roosevelt administration and flamboyant personality also garnered him his share of critics. Pinchot left Washington to become governor of Pennsylvania, but he maintained an active voice in national forest matters until his death in 1948. To replace Pinchot, Taft selected Henry Graves, former assistant to Pinchot and current dean of Yale's Forestry School.

However, the transition to a new chief forester was not the only event that marked the year 1910 as a turning point for the national forest system. It also witnessed the creation of the new Forest Products Laboratory at the University of Wisconsin. Emphasizing the development of new timber products as well as logging and milling technologies, the Laboratory underscored the Forest Service's commitment to commodity production and tree farming on national forests.

But perhaps most significantly, the year recorded a series of some of the most catastrophic forest fires in the nation's history. The worst of these was labeled the "Big Blowup" or "Big Burn," which took place in the panhandle of Idaho. According to author Timothy Egan, the fire claimed almost one hundred lives and attracted massive media attention.[18] Newspaper accounts at the time glorified the patriotic and virtuous qualities of the foresters who died trying to fight the blaze. Their popularity was further heightened with the release of Zane Grey's new novel, *The Forester*, in which a valiant, hardworking, and patriotic member of the U.S. Forest Service was the hero.

The Big Burn came on the heels of other fire-related disasters, including a devastating flood on the Monongahela River in West Virginia in 1907. Because forest fires had stripped away trees and vegetation, there was no longer any means of slowing runoff after heavy rains. Floodwaters inflicted over $100 million in damage as they scoured agricultural lands before finally slamming into the city of Pittsburgh, leaving death and ruined neighborhoods in their wake.

These catastrophic events were enough to turn public opinion in favor of the Forest Service and put sufficient pressure on Congress to pass the Weeks Act of 1911. The Act was instrumental in shaping the national forest system in two ways. First, it helped cement a new Forest Service relationship with state agencies and private timber interests around a two-pronged fire policy of prevention and suppression. Second, it used this new concern with forest fires as a rationale for expanding the national forest system into the eastern United States.

Specifically, the Weeks Act granted federal matching funds to states to create new Forest Protection Agencies designed to fight and prevent wildfires. Additionally, at the request of eastern states, the Act provided funds for the purchase of private lands as new national forests. In many cases, these lands had been previously logged or otherwise denuded. As national forest lands, federal managers sought to restore them in order to protect watersheds, prevent flooding, and prohibit future fire events. To avoid Supreme Court challenges, the Weeks Act contained very precise wording. The stated purpose of land acquisition was to protect "watersheds of navigable streams," presumably to aid commercial transportation activities. In this way, the law aligned with Congress's constitutional authority to intercede in matters of interstate commerce.

Pinchot and others had long pushed to expand the national forest system in the East, but the idea never gained much traction in Congress. But now with the goal of fire and flood protection front and center, the rationale for eastward expansion garnered broad congressional support. In 1916, the Pisgah National Forest in North Carolina became the first national forest created in the eastern United States. This was followed by White Mountain National Forest in New Hampshire and George Washington National Forest in the Southern Appalachians of West Virginia and Virginia.

In 1924, the Clarke-McNary Act brought the eastern forests into closer alignment with those in the rest of the system by declaring watershed protection *and* timber production as their primary purpose. It also required states to provide matching funds for the federal purchase of private lands, and it provided incentives for landowners to work with the Forest Service to improve the health of their privately owned forest lands.

As the United States entered the Great Depression, an enormous amount of additional eastern forestland was put up for sale at bargain prices. Unlike land purchases for national parks, which tended to be aesthetically pleasing (and therefore more costly), the land purchased for national forests was generally in poor health. Much of it had already been cut over through unsustainable logging methods, leaving behind seriously eroded landscapes. Many landowners wished to sell, and given the dire economic climate, the federal government was one of the few buyers on the market. As a result, land often sold for less than $5 per acre. Once the land was purchased, the Forest Service went to work on replanting and rehabilitation projects, with ample assistance from the Civilian Conservation Corps (CCC).

These efforts were given additional financial support in 1930 with the passage of the Knutsen-Vandenberg Act. This law directed funds earned through federal timber sales to be used specifically for reforestation programs. In a matter of decades, these new national forests began to transform once again into forested landscapes. Historian James Lewis notes that by 1961, almost twenty-one million acres of new national forest were added to the system in this way.[19]

Yet despite the Forest Service's emphasis on fire suppression during this period, forest fires continued to be numerous and severe. The Forest Service benefitted from the additional manpower provided by CCC workers to build firebreaks. The agency also deployed new technologies, including airplanes and smokejumpers, and it adopted new stringent policies such as the so-called 10 a.m. policy, which declared that every reported fire be suppressed by 10:00 a.m. the next day. Nonetheless, over two hundred thousand forest and range fires took place each year throughout the 1930s. The Forest Service and other groups recognized a need for greater fire *prevention* efforts (not to be confused with *suppression*), including programs to educate the general public about the causes of wildfire. Though fire prevention programs had existed for years, the issue grew to vital national importance during World War II (see textbox 5.1). A partnership soon emerged that led to a public advertising campaign featuring a large, talking black bear named Smokey (figures 5.4a, 5.4b, and 5.4c).

Figures 5.4a. The first image of Smokey Bear appeared in 1944. Image courtesy of the National Archives and Records Administration, used with permission of the National Ad Council and USDA Forest Service.

Figure 5.4b. A few years later, Smokey adopted the famous catchphrase, "Only you can prevent forest fires," and took on the appearance familiar to millions as depicted in this 1951 poster designed by artist James Hansen. Image courtesy of the National Archives and Records Administration, used with permission of the National Ad Council and USDA Forest Service.

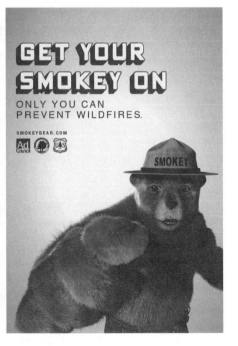

Figure 5.4c. A more recent manifestation of Smokey Bear derives from a computer-generated image linked to a wide range of social media outlets. Image courtesy of the National Ad Council and the USDA Forest Service.

TEXTBOX 5.1.
The Story of Smokey Bear

Though immediately recognizable to millions of Americans, few may realize that Smokey Bear was invented by the Ad Council, the Association of State Foresters, and the USDA Forest Service as part of the war effort during World War II. Amid numerous campaigns to conserve natural resources, timber began to receive special attention when Japan shelled an oil refinery near the Los Padres National Forest in California and federal officials realized that forest fires could be started in a new way. Additionally, between 1944 and 1945, over nine thousand incendiary balloon bombs were launched from Japan or released by Japanese warships. Though only a handful succeeded in starting fires, the threat they posed provided continuous pressure for government action. Yet even before the first incendiary bombs were launched, efforts were well under way to develop a new public awareness campaign for the prevention of forest fires in the United States. In August 1944, after experimenting with a number of animal figures including Bambi from the Disney movie, Smokey Bear was finally chosen for his friendly yet noble and sincere expression. In 1947, Smokey adopted his famous catchphrase, "Only you can prevent forest fires."

In 1950, a badly burned bear cub was rescued from a wildfire in New Mexico and brought to the National Zoo in Washington, D.C., to serve as the physical embodiment of Smokey. Two years later, a stuffed bear in jeans and hat was released for sale as a children's toy. After a "junior ranger" membership card was made freely available to the public, over half a million youngsters sent in their cards for registration. In 1964, Smokey's popularity was so great that he received more mail than the president. On April 29 of that year, Smokey earned his own zip code from the Postal Service. Though the living Smokey died on November 9, 1976, the popularity of the Smokey Bear icon persists. A recent campaign advertisement includes a CGI-produced image with the catchphrase, "Get your Smokey on!" Smokey represents the longest-running public ad campaign in American history. And according to historian James Lewis, Smokey remains the third most recognized popular culture icon in America, after Santa Claus and Mickey Mouse.

THE IDEA OF MULTIPLE USE

In the beginning of this chapter, we noted how the challenge of balancing multiple resource uses not only distinguished national forests from the dualistic mission of the national parks but also explained many of the persistent conflicts—both conceptual and political—faced by Forest Service managers. While timber production and the protection of water supplies provided the rationale for establishing the national forest system in the late 1880s, and fire suppression policies helped to justify its expansion at the turn of the nineteenth century, a more formal and complex expression of the multiple-use approach to management did not emerge until subsequent decades. This shift was catalyzed by a number of corresponding forces, including increased demand for timber, debates over livestock grazing fees, and calls for greater recreational opportunities on public lands.

During the Forest Service's first three decades, the timber harvest on the national forests was relatively small, amounting to less than one billion board feet per year. This custodial approach to management encouraged the use of selective logging techniques and was designed to reserve the nation's forests for times of crisis. Pinchot insisted that a future "timber famine" in the United States was imminent. He believed that conventional industrial logging methods were harmful to the nation's forests, and if allowed to continue, they would ultimately deplete the resource. But there was another, perhaps even more important, rationale for minimal harvests on the national forests during this period. Namely, it was what the private timber industry wanted. At the behest of timber company executives, the Forest Service agreed to avoid flooding the market with public timber that might drive down prices.

However, with the onset of war in the early twentieth century, everything changed. During World War I and continuing into the dramatic industrial growth of the Roaring Twenties, logging boomed on private forestlands. The national forests also began to feel pressure to increase timber production. Unfortunately, according to Pinchot, the boom was fueled in part by the implementation of harmful but economically desirable logging practices such as clear-cutting.

But this heightened pace of timber production would not last. As the Depression Era dawned, the heightened demand for wood products quickly evaporated. In 1933, the Forest Service took the opportunity to issue the Copeland Report, a comprehensive assessment of the national forest system. The report called for an expansion of the system and greater regulation on privately owned and managed forests. That same year, Bob Marshall published his book *The People's Forests*, in which he went so far as to call for the nationalization of *all* forestlands in the United States. Though some Copeland Report recommendations were eventually adopted via FDR's New Deal conservation programs, calls for tighter regulatory control of private and public forests were quickly silenced when the United States entered World War II. Between 1940 and 1941, national consumption of lumber grew from five billion board feet per year to twelve billion. For national forests, this signaled the beginning of a steady increase in timber harvest levels that would continue into the early 1990s.

After the war, the demand for timber production intensified even more. Economic prosperity in the United States coupled with the need to rebuild a war-torn Europe fueled ever-growing markets for lumber. As private supplies began to dwindle, pressure mounted for increased production from the national forests. In 1952, the Dwight Eisenhower administration and the newly appointed Forest Service chief, Richard McArdle, enthusiastically embraced this new management priority. As a result, national forest harvests ballooned from 5.6 billion board feet per year in 1950 to 12.8 billion in 1969.[20]

Some in the Forest Service welcomed this change, believing the increase in production might finally allow the agency to meet the elusive goal of financial independence.[21] However, doing so meant logging previously untouched stands, which, ironically, required an *increase* in congressional funding to pay for new logging roads. As Zaslowsky and Watkins point out, this soon became an exercise in circular logic. Proceeds generated from expanded logging were off-set by new road costs. Covering these required additional logging, which required additional roads, and so on.[22] And as noted earlier, the requirement to hand over 25 percent of receipts from timber sales to local government coffers not only presented another incentive to increase production, but simultaneously placed the goal of balancing the books that much further beyond reach. Meanwhile, all of this road building was creating a new, as yet unacknowledged, management problem for the national forests. By 1989, the system of logging roads extended over 360,000 miles, making it one of the most extensive in the world.[23] This situation presented new sets of challenges. How should the Forest Service manage these roads, when the two options available, maintenance or environmental restoration, both cost money?

Amid this upsurge in timber production, questions began to arise regarding the suitability of adhering to multiple use as the guiding principle for national forest management. At the very moment timber production ramped up, so did demands on the national forests for other competing uses. It was one thing to espouse a multiple-use management philosophy when in fact a single use—timber production—was predominant, and was only lightly implemented at that. From the agency's inception until World War II, this was generally how the Forest Service operated. But it was quite different when multiple, diverse, and incompatible interests began to clamor aggressively for ever larger slices of a finite national forest pie.

Pressure to expand the types of resource development on the national forests emanated from a variety of sources. From the beginning, mineral development and livestock grazing were identified as legitimate uses of national forests. Nonetheless, many in the western livestock industry were still disgruntled with the Forest Service over the imposition of grazing fees in 1906. Others were angry and bitter from more recent events, including the cattle industry's failed attempt to block FDR's declaration of Grand Teton National Monument near Jackson Hole, Wyoming, in 1943. However, in 1946, a new surge of antipathy toward the Forest Service was triggered by a proposal to reduce grazing numbers on permits in order to allow forestlands to regenerate after heavy wartime use. In response, western livestock interests convened in Utah to launch the first Sagebrush Rebellion. The stated goal was nothing less than the privatization of all western public lands, but the rebels' approach was a bit more subtle. Ostensibly, they framed the issue as a question of states' rights and proposed the transfer of all federal grazing lands—both national forest and BLM land—to the states in keeping with their presumed constitutional rights. However, once in state control, they expected local officials to sell the lands to private landowners (e.g., the ranchers who held grazing permits) for the negligible price of nine cents per acre. Learning of the plan, author Bernard DeVoto wrote a scathing critique in a series of essays in *Harper's Magazine*, bringing to an end any serious consideration of the idea in Congress. Although the threat of the Sagebrush Rebellion was temporarily defused, discontent among livestock growers remained and would erupt several more times in the coming decades (see chapter 7).

Meanwhile, an even greater test of the multiple-use philosophy appeared in the guise of recreation and wilderness protection. Gifford Pinchot had been clear since the beginning that recreational activities and aesthetic values on national forests were merely incidental. He drew a sharp line between the idea of conservation and the concept of preservation that John Muir so eloquently advocated as a guiding philosophy for the national parks. Their very public dispute over efforts to dam Hetch Hetchy Valley underscored these differences. And though Pinchot's view ultimately won out, it also tarnished the reputation of the Forest Service in the eyes of the public. The goodwill engendered after the big fires of 1910 diminished somewhat after passage of the Hetch Hetchy dam project and John Muir's death in 1914.[24]

Tensions between the national forests and parks, however, were not only philosophical but also carried significant material implications. In the same way that park supporters mourned the loss of Hetch Hetchy in Yosemite National Park to development, national forest advocates began to decry the loss of forestland to the Park Service for purposes of preservation. From the late 1890s onward, it was not uncommon for new national parks to be carved out, at least in part, from national forestland.[25]

In response, the Forest Service fought vigorously against Stephen Mather's efforts to create a National Park Service in 1915. Arguing that recreation was one of many multiple uses already administered by the Forest Service, they claimed a Park Service would be redundant. It is perhaps no accident that the Forest Service enacted several new initiatives during this period to establish its own recreation credentials, including a new public relations program touting the national forests as the "People's Playgrounds." In 1915, a new permit program allowed individuals to lease national forestland in remote areas to build summer cottages, lodges, and businesses for recreational purposes. The next year, the very first national forest campground was created in Oregon's Mount Hood National Forest. And in 1924, at the urging of Aldo Leopold, the Forest Service established a new land management designation on national forests called "primitive areas"—places devoted to wilderness and wildlife habitat protection (see chapter 8).

But it was too little, too late. Stephen Mather was to the Park Service what Pinchot had been to the Forest Service: a man accustomed to, and skilled in, the art of political persuasion. Using Pinchot's arguments against him, Mather successfully argued that conservation and preservation were *not* the same thing, and that the national parks needed their own management agency.

Despite the fact that the Forest Service "lost" the battle for recreation to the National Park Service, pressure by recreational users on national forests increased dramatically in the postwar era. The same prosperity that drove the growing demand for lumber (and roads) in the national forests also fueled demand for recreation opportunities. The way the public viewed, valued, and used the national forests was changing. With completion of the newly minted National Highway System and expanded opportunities for leisure time among the middle class, a growing number of visitors found their way to national parks *and* national forests. Prewar estimates of national forest visitation number approximately ten million per year. By 1960, the numbers skyrocketed to 190 million annually.

But matters were coming to a head for Forest Service personnel. The growth and intensification of conflicting demands for timber, livestock grazing, mineral development, and now, recreation, stretched the multiple-use philosophy to the limit. With the national forests primed for crisis, the time had come to revisit the concept of multiple use in the hope of providing guidance for managers.

To this end, Congress passed the Multiple Use and Sustained Yield Act in 1960. If one could clarify multiple use as a management goal, perhaps it would help dispel some of the confusion over management priorities and defuse conflicts between the Forest Service and resource users. Not surprisingly, criticism sprang from all sides. For example, the National Lumber Manufacturers Association opposed the bill fearing that it would water down what it saw as the original rationale for the national forests under the 1897 Forest Organic Act; namely, protecting watersheds and providing the nation with a continuous supply of timber. Conversely, the

Sierra Club feared the law would merely legitimize the agency's historical priority of logging over other forest values since Forest Service personnel held no training in other fields of conservation. Though it eventually acquiesced, the Sierra Club withheld final endorsement until the bill was amended to include wilderness protection as one of the sanctioned "multiple uses" of national forests.

In its final form, the Act identified five major uses of national forests, all deemed to be of equal importance. Listed in alphabetical order to avoid any sense of favoritism, they included: outdoor recreation (including wilderness protection), range (or livestock grazing), timber production, watershed protection, and the provision of wildlife and fish habitat. Though not listed as one of the five priorities, mineral development maintained its unique status as a "highest and best use" on national forestlands. In addition, the "sustained yield" portion of the law codified the practice that timber harvests would never exceed the amount of wood grown in a given year in order to ensure the sustainability of the system. In hindsight, the law did little to change actual management practices since its interpretation was often left up to local managers on individual forests. Just as the Sierra Club feared, timber production, followed by grazing and mining activity, tended to dominate management priorities on the national forest system for the next few decades.

CLEAR-CUTTING, NFMA, AND BELOW-COST TIMBER SALES

As the Environmental Decade dawned in the mid-1960s, things did not get any easier for the Forest Service. As Congress passed a host of new natural resource protection laws, environmental NGOs gained new popularity, and environmental awareness soared among the general public. In contrast, the steady escalation in timber harvest levels on national forests appeared increasingly out of step with these new sentiments. To make matters worse, the Forest Service chose this moment to expand the use of clear-cutting as the preferred method for logging. Environmental activists and the general public took notice of these developments, and did not like what they saw.

The environmental laws passed during the decade provided new legal tools for environmental groups to express their concerns. Among the new statutes, environmental activists found the 1969 National Environmental Policy Act (NEPA) to be particularly effective. By requiring federal agencies to provide environmental impact statements (EISs) for all management projects, and to include public participation in the process, NEPA ensured federal managers would face greater public scrutiny over management decisions.

Unfortunately, the Forest Service was not always prepared for this scrutiny. Since its founding, the agency was charged to manage the nation's forests in such a way as to benefit the public good. To a large degree, Forest Service authority and, indeed, its institutional legitimacy, rested heavily upon its ability to apply scientific expertise and unbiased judgment to this task. In this context, the mere thought of muddying the waters with the viewpoints and opinions of untrained laypersons seeking to advance their own agendas seemed at odds with the mission of the agency. Of course, some citizen activists were also scientists in their own right, ready to offer local knowledge that might challenge, and potentially improve, Forest Service decisions. While some officials welcomed public participation, others felt that it served little purpose other than to undermine the scientific authority of Forest Service managers.

But of course, public participation under NEPA entailed communication in both directions. In addition to collecting public comments, federal agencies were also required to explain land use decisions to the general public. Fulfilling this requirement once again posed a significant challenge for the Forest Service. This was due in part to the problem of translating into plain language the acronym-laden jargon and complex terminology of professional forestry. It also reflected the sheer difficulty of explaining the scientific nuances of complex topics in forest ecology. In such situations, public meetings could readily transform into a minefield of political missteps.

Consider the case of fire policy. Protecting the nation's forests from wildfire and wasteful industrial logging techniques represented two of the major rationales for creating the national forest system in the first place. For the general public in the early 1970s, forest fires constituted a clear and present danger. Very few understood the distinctions between massive wildfires and fire events taking place on smaller scales or in diverse vegetative or climatic conditions. Forest managers thus faced a monumental task in explaining how fire could be a positive thing. Depending on specific intensities and frequencies, regular small-scale fire events are a necessary component in many forest landscapes, helping to maintain ecological health and to *reduce* the threat of large-scale outbreaks. But in the presence of Smokey Bear, this was a difficult case to make, especially to those who struggled to discern the difference between fire *prevention* and *suppression*.

In a similar vein, clear-cutting represented one of the most abusive industrial logging practices of the past—the type of environmental mistreatment that the Forest Service was supposed to guard against. However, in certain circumstances, when particular species and landscapes were present, clear-cutting could be the best choice for replicating natural forms of regeneration. For example, aspen stands reproduce themselves largely from the existing root system and do so most vigorously when the entire stand is brought down from catastrophic events. Such events might be caused by fire or windstorms, but they could also be simulated via clear-cutting.

On the other hand, when used indiscriminately, clear-cutting is also capable of producing significant environmental damage. Deployed in areas with unsuitable tree species or on steep hillsides, clear-cutting can radically degrade forest landscapes—destroying wildlife habitat, allowing massive erosion, and creating conditions for large-scale wildfires, invasions of exotic species, and flooding. As mentioned above, such irresponsible practices provided the rationale for the creation of the national forest system back in the 1890s. For many environmental groups, scientists, and the general public, the Forest Service's broad adoption of clear-cutting in all kinds of forest environments, simply in order to "get out the cut," suggested the worst kind of hypocrisy.

In response to all of this, Congress passed the 1976 National Forest Management Act (NFMA). While the Act did not outlaw clear-cutting altogether, it did establish strict guidelines for its use that radically limited future applications. The law also called for the creation of long-term (fifteen-year) comprehensive management plans for each national forest in the United States. With NEPA and NFMA now in place, environmental groups had the means to challenge not only individual timber sales but also long-range management plans on issues ranging from timber production and livestock grazing to recreational activities.

In the 1980s, the trend toward increased timber harvest levels accelerated even more. Environmental groups and even some Forest Service managers accused the Reagan administration of seeking to liquidate the national forest timber supply in order to drive up prices on private timberlands. Though clear-cutting was now in check, the new legal strategies provided in NEPA and NFMA allowed activists to turn their attention to other questionable management practices, including below-cost timber sales.

At issue was the cost charged to private loggers and mill operators who subcontracted with the Forest Service to cut timber on national forestlands. As stipulated in the 1897 Forest Organic Act, logging on national forests was not carried out directly by the Forest Service. Rather, private businesses submitted bids for timber sales arranged and managed by the agency. However, environmental groups charged that the prices charged by the Forest Service for public timber were not only below market price but also below the costs of preparing the sale. These costs included such things as surveying timber stands to determine the boundaries of the sale, building roads to the site, marking trees to be cut or left behind, drawing up the paperwork, and finally, monitoring the logging operation and any required restoration work taking place afterward. By failing to recover these costs, critics charged that the Forest Service timber program amounted to nothing more than a subsidy to the private logging industry funded by taxpayers. Environmentalists leveled similar charges against the federal livestock grazing programs administered by the Forest Service and the Bureau of Land Management, in which, once again, fees failed to match fair market value of the forage provided on public lands (see chapter 7).

The Forest Service and timber industry responded that the mismatch between costs and revenue resulted from the higher regulatory standards found on public forestlands. Maintaining these standards required more paperwork, increased monitoring activity, and greater restoration efforts than one typically finds on private lands. Moreover, despite efforts by the Forest Service to be financially independent (pay for its own programs via timber and other use receipts), the primary mission of the agency was not to turn a profit. Rather, the Forest Service was also mandated to provide economic support to local rural communities. Insofar as logging companies created local jobs, these programs were justified. While these arguments failed to convince critics, by the end of the decade a new dispute appeared on the horizon that quickly eclipsed the issue of below-cost timber sales. Before long, the entire nation turned its attention to the fate of a rare species of raptor that lived in a singularly unique forest habitat: the northern spotted owl.

CONFLICT SOARS TO NEW HEIGHTS:
THE NORTHERN SPOTTED OWL

For many observers, the dispute in the Pacific Northwest over the fate of the northern spotted owl represents the pinnacle of national forest conflict in the United States. It helped to shape the contours of environmental politics for decades to come. The logging industry and the burgeoning Wise Use Movement effectively defined the issue as one of "jobs versus the environment." From this perspective, environmental issues constituted a zero-sum game. In a region where the local economy was so bound to the timber industry, environmental regulations could only come at the expense of human welfare. Though it ignored the political, economic, and technological complexities of the issue, this way of framing the debate profoundly influenced popular media coverage.

On the other side of the debate, environmental NGOs and forest ecologists argued that restrictions on logging were needed to protect threatened or endangered species as well as one of the most unique and valuable ecosystems on the planet: old growth temperate rain forests. They noted, moreover, that both the science and the law were on their side. Realizing that these views held little currency with local rural residents in the Pacific Northwest, environmental activists successfully turned the campaign into an issue of national concern.[26] By the time it was over, after years of political, legal, and verbal wrangling, the conflict significantly transformed both the environmental movement and national forest management in the United States, opening the door to new collaborative forms of conservation.

At the center of the conflict was the question of listing the northern spotted owl as either "threatened" or "endangered" under the 1973 Endangered Species Act (ESA). If the U.S. Fish and Wildlife Service (USFWS) decided to list the owl, the ESA required that steps be taken to protect the owl's habitat. This meant removing potentially thousands of acres of old growth forests from the possibility of commercial logging. Since the late 1960s, ecologists recognized that both the owl and its old growth forest habitat were at risk by Forest Service harvesting practices. However, it took several decades more to gather sufficient evidence and convince federal managers to take action on the issue. After passage of NFMA in 1976, the Forest Service attempted to incorporate northern spotted owl protections into its long-term comprehensive management plan for the region. But the lack of substantial actions to preserve the owl habitat resulted in a backlash from environmental groups.

In 1986, an obscure environmental organization in Massachusetts filed a request with the USFWS to list the owl as endangered under the ESA. Under intense pressure from the Reagan administration, the regional USFWS director turned against the advice of his own wildlife biologists and recommended that the owl not receive listing. The Sierra Club Legal Defense Fund filed a lawsuit in federal court, which ruled against the USFWS, ordering them to reconsider the evidence for decision. In 1989, a study by the U.S. General Accounting Office reaffirmed the fact that the USFWS did not act in good faith.[27]

Realizing the need for external support amid decidedly hostile local logging communities, environmental groups launched a nationwide campaign to protect the owl, writing articles and garnering national media attention in outlets such as the *New York Times* and *National Geographic*. While mainstream groups led tours and organized congressional lobbying strategies, radical groups such as Earth First! staged acts of civil disobedience: setting themselves up in forest stands slated for logging, chaining themselves to trees, or committing acts of vandalism on logging equipment and vehicles.

Local logging interest also looked further afield for support, receiving backing from international timber corporations and property rights advocacy groups, including those who championed the idea of "wise use."

Borrowing the phrase from Gifford Pinchot, members of the Wise Use movement effectively framed the debate in terms of jobs versus the environment. In so doing, they offered a powerful populist message blending resource development on public lands with the goals of democracy and social justice for rural working families. In contrast, they presented those who valued nature in nonutilitarian terms (environmentalists) as elitists who cared nothing for "common" people. These ideas set the rhetorical groundwork for federal deregulatory campaigns mounted by conservative politicians for years to come.

Finally, in June 1990, the USFWS reversed its original decision and opted to list the northern spotted owl as "threatened" under the ESA. This move set off a fresh round of legal fights over how much old growth forest should be set aside as critical habitat for the owls. In 1992, the Clinton administration attempted to settle the matter by convening the Northwest Timber Summit. The resultant Northwest Forest Plan of 1994 sought to balance the protection of the owl habitat with the continuation of limited logging activity, while aiding local timber workers by subsidizing job-retraining programs. Over the years, the specifics of the Northwest Plan have been continually challenged and revised. Between the 1995 Timber Salvage Rider and the 2003 Healthy Forest Restoration Act (see below), environmental activists faced a constant barrage of efforts to weaken protections for the owl.[28]

However, the most lasting effect of the Northwest Forest Plan may have been the blueprint it provided for ecosystem-based management and collaborative forest planning.[29] While the Plan was clearly a top-down initiative, it convened a diverse set of stakeholders in the decision-making process. Representatives from the logging industry, environmental NGOs, local government, and managers from a diverse set of federal agencies, including the Forest Service, the Bureau of Land Management, and USFWS, all worked in a collaborative manner to craft an ecosystem-wide management plan. This represented an historical departure from the Progressive Era–principle of a single federal agency issuing decisions based purely upon its own scientific expertise. And by adopting an ecosystem perspective to determine management strategies and priorities, managers changed both the process and criteria used for public land management and planning.

Other examples of collaborative ecosystem-based management soon followed, including the Interior Columbia Basin Ecosystem Management Project (ICBEMP) and the Sierra Nevada Ecosystem Project (SNEP). These agency-initiated collaborative planning projects were soon matched with a growing number of *grassroots*-initiated, community-based collaborative approaches to public land management. In efforts such as the Quincy Library Group in California, the Applegate Partnership in Oregon, or the

Ponderosa Pine Forest Partnership in Colorado, local residents looked for new collaborative ways to escape the legacy of nineteenth-century ideas and stereotypes that seemed to prohibit efforts to resolve difficult land use problems on national forests (see case studies below). By working together and engaging in frank, open discussions, participants might find opportunities to break away from static notions of nature as "empty" and pristine, and instead give greater attention to the needs and expectations of local communities in management planning. While critics of collaboration questioned all such claims, others held out the hope that it might provide a means of moving beyond the limits of narrow nature-as-commodity thinking: to consider other forms of value in nature, and to acknowledge the inherent dynamism in ecological systems, including that rendered by fire.

THE HEALTHY FOREST RESTORATION ACT OF 2003

If the northern spotted owl conflict triggered new thinking in logging practices and management decision making for the national forest system, the next major turning point was brought about by the issue of wildfire. The summer of 2000 witnessed some of the largest wildfires on record in the American West. Over 2.3 million acres of national forest-land burned that year, the most since 1919.[30] While the Yellowstone fire in 1988 drew public attention to the issue, large-scale events in 1994 (in which over a million acres burned) and again in 1998 added to the feeling that something was amiss in national forest management. Over the next decade, massive fires broke out throughout the Mountain West, setting new records for acreage burned and houses or other buildings destroyed.

As noted earlier, since its founding, the Forest Service had dutifully carried out a policy of fire suppression on the national forests under the dual rationale of protecting the public and preventing resource waste. By the 1930s, the agency adopted a "fires out by 10:00 a.m. policy," employed airplanes and "fire jumpers," and used new flame-retardant chemicals. By the 1940s, the Forest Service had adopted a fire *prevention* campaign led by Smokey Bear to complement its fire suppression policy (see textbox 5.1). But by removing all fire from the environment, including regular small-scale events, the policy produced unexpected ecological effects.

Most western forests were adapted to fire. In fact, some species depended upon it to carry out vital ecosystem functions. In ponderosa pine forests, for example, low-intensity fires every five to seven years play an important role in clearing out underbrush, thinning young tree stands, and fertilizing soils. While the outer bark at the base of the pines might become charred, mature trees are otherwise unharmed and the upper branches unburned. In other forest types, fire renders different effects and tends to occur at different time intervals. Species such as the lodgepole and jack pine have "serotinous cones": cones covered in resin that depend on heat (often from forest fires) to melt the resin, open the cones, and release the seeds.

Decades of fire suppression, however, allowed the underbrush in many western forests to grow dense and reach heights equal to the lower branches of mature trees. Under such conditions, when fires do break out, the effects can be catastrophic. With stockpiled fuel, fires burn hotter, reaching temperatures that can destroy organisms in the soil. And as the flames reach tree branches, they may transform into "crown fires," burning entire trees and triggering massive events. In combination with the effects of past harvesting practices such as clearcutting, which allowed forest stands to grow back with trees of similar age, size, and species, forests were not only at increased risk of fire but also of disease and insect outbreaks.[31]

In order to address this problem, in the early 1970s the Forest Service began to change its fire policy to allow some naturally caused fires (most often from lightning strikes) to burn themselves out. The agency also began to consider the prospect of introducing controlled burns. However, these ideas ran into intense public opposition after the 1988 fires in Yellowstone, which began as a fire that the agency had decided not to suppress.

In response to these concerns, the Clinton administration initiated investigations in the early 1990s, which culminated in the 1995 Federal Wildfire Management Policy. In 1999, Congress commissioned its own study sponsored by the Government Accountability Office on ways to address the threat of catastrophic fires. The culmination of these reports led to the National Fire Plan of 2000. The plan called for the reduction of hazardous fuels through logging activities aimed at thinning the forests. It continued to support firefighting policies and the restoration of burned landscapes, but it also ordered the Forest Service to work closely with local communities to develop individualized plans to reduce fire risk.

In 2003, Congress weighed in with the Healthy Forest Restoration Act, endorsed and signed by President Bush. The law allowed selective logging on national forests in order to thin forest stands. It then called for the reintroduction of fire via controlled burns. In this way, forest managers hoped to mimic natural fire regimes, thereby reducing the risk of massive fire or disease events.

For critics, however, the name of the law belied its true meaning. Ecologists and leaders of mainstream environmental NGOs agreed that some forest thinning and fire reintroduction was necessary to improve the health of national forests. However, they preferred to focus these treatments in buffer zones around mountain towns and in places where human residential subdivisions blended into forest environments, areas referred to as the wildland-urban-interface (WUI). Due to the enormous expense of selective logging and controlled burns, a targeted approach made much more economic, logistical, and ecological sense. With private property thus protected, nature could take its own course elsewhere on the national forests.

Environmental groups took issue, therefore, with the fact that the Healthy Forest Act called for thinning and fire reintroduction on the *entire* national forest system. Moreover, to help cover the costs of this endeavor, the Act allowed loggers to cut a number of economically valuable large-diameter trees along with the smaller-diameter trees needed for thinning. In addition, the law expedited restoration operations, allowing them to bypass certain regulatory standards required for conventional timber sales. At the same time, and rather significantly, it opened up places heretofore protected from logging activity, including Roadless Areas, endangered species habitats, and wilderness study areas. In short, opponents feared that—in the name of wildfire prevention—the Healthy Forest Act would allow timber interests to circumvent forest protection laws and abuse its provisions to cut commercially valuable large-diameter trees in protected areas.

In practice, environmental groups tended to be relatively successful in blocking thinning projects slated for sensitive areas under the law. And for the Forest Service, efforts to thin forest stands proved to be challenging at best. Even with the provision to cut some large-diameter trees, the costs of selectively logging smaller trees continue to defy profitability in many places across the West. As a consequence, it can be difficult to secure timber contracts for these types of sales. Meanwhile, efforts to reintroduce fire through controlled burning have presented an even greater problem. Not only is it difficult to implement prescribed burns when weather conditions are right (it cannot be too wet, too dry, too hot, or too windy), but the sheer amount of forest in need of such treatment is simply overwhelming. As a consequence, over the past decade, much of the restoration work under the Healthy Forest Act has, in fact, focused in the WUI, while massive-scale wildfires have continued to grab national headlines.

COMING FULL CIRCLE

Over the past one hundred years, the way in which we have come to understand, use, and value the national forest system has evolved in significant ways. Measured in terms of total annual timber harvests,

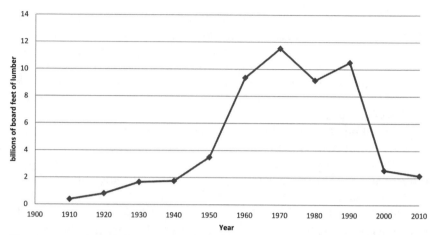

Figure 5.5.   Annual Timber Harvests on National Forests at Ten Year Intervals, 1910 to 2000. From 1962 to 1980 and again from 1983 to 1990 annual harvests totaled between 9 and 12.7 billion board feet each year. Though not shown here, the year 1987 marked the highest harvest level with 12.7 billion board feet. *Source:* **USDA Forest Service, "FY 1905– 2012 National Summary Cut and Sold Data," http://www.fs.fed.us/forestmanagement/ products/sold-harvest/index.shtml.**

management priorities appear to have come full circle. As figure 5.5 illustrates, the custodial approach that dominated national forest management from the 1890s until World War II rendered relatively small annual harvests. Then, in the postwar era, and especially in the three decades from 1960 to 1990, the intensification of timber production maximized commodity values and pushed harvest levels to new heights. However, in the aftermath of the northern spotted owl conflict and the increasing influence of region-specific ecosystem management principles in the early 1990s, timber harvests returned once again to levels not seen since the dawn of the nineteenth century.

Of course, reasons for low timber harvests differ between the two eras, but both are products, at least in part, of dominant nineteenth-century ideas of nature. One reason for reduced annual harvests today is that most of the large, easily accessible, and commercially valuable trees on national forests were cut during the logging heyday of the 1960s, 1970s, and 1980s. Arguably, the notion of nature-as-commodity was a major driver for expanded production during this time. In contrast, the trees currently dominating national forest landscapes are simply too young and too small to justify—or even support—the economics of large-scale logging.

However, another reason for reduced harvests is the rise of what Samuel Hays calls ecological forestry, in which forest managers emphasize noncommodity resource values (as well as nontimber forest products), including forest restoration, wildlife habitat, and biodiversity, while also giving greater attention to recreational and aesthetic concerns.[32] Yet one of the rationales for this turn toward forest restoration was to address the negative environmental impacts rendered by adherence to the agency's fire suppression policy; one that was rooted in the conception of nature as unpeopled, eternal, and pristine. In other words, fire policy embodied the idea that a forest preserved and protected from fire would stay preserved and protected (albeit with larger trees). By preserving the status quo, the nation's forests would remain healthy and available for logging when future need arose.

Managerial changes were also a product of changing global economic conditions, including the availability of inexpensive timber resources in other countries, relatively cheap transportation costs, as well as changing societal values over the preferred use and management of public lands and resources. But despite this complex array of causal forces for changing management priorities, the impacts of nineteenth-century conceptualizations of nature remain palpable.

CASES

The continued legacy of these ideas is illustrated below in two case studies that document recent attempts to further reshape forest management priorities. The first case explores the rise of two collaborative conservation efforts in national forest management as demonstrated by the Quincy Library Group in California and the Ponderosa Pine Forest Partnership in Colorado. As some of the first collaborative efforts of their kind, the initiatives point the way toward an alternative approach to grappling with diverse sets of public land uses and values, potentially breaking free of some of the dominant conceptions of nature noted above. In the second case, debates over the Roadless Area Conservation Rule speak to the challenges of implementing environmental preservation as part of the Forest Service's multiple-use mandate. The long, drawn-out conflict over this administrative rule change underscores the continued legacy of nature-as-commodity thinking and harkens back to some of the same conceptual tensions illustrated in the founding of Yellowstone National Park.

From Quincy to Cortez: Collaborative Conservation and National Forests

The Quincy Library Group represents what is perhaps the most well known collaborative effort in the nation. Located in a small logging town (population five thousand) in California's Sierra Nevada Mountains north of Lake Tahoe, the group famously developed an alternative timber management plan covering 1.53 million acres across three national forests: the Lassen, Plumas, and Tahoe. Unprecedented both for its audacious scope and unique decision-making approach, the plan garnered national headlines when Congress took up the issue and overwhelmingly passed a law forcing its adoption by the Forest Service in 1998.[33]

The effort began in the 1980s. In the midst of massive annual timber cuts, as logging trucks loaded down with giant trees constantly rumbled out of the national forests, a local environmental group called the Friends of the Plumas Wilderness decided to take action. In 1986, they proposed an alternative timber plan that would protect all remaining roadless areas and focus logging on roaded second-growth forests using methods that would thin the stands rather than use large-scale (forty-acre) clear-cutting techniques. Forest Service officials ignored the plan in favor of business as usual. Within a few years, however, the situation had changed dramatically.

By the early 1990s, years of heavy logging had severely depleted the remaining inventory of large and easily accessible trees. Decades of fire suppression, clear-cutting, and even-age management produced a forest landscape filled with densely packed, small-diameter trees with little commercial value other than for pulp wood products. The writing was already on the wall when, in the wake of the northern spotted owl controversy up north, the Forest Service took steps to protect the owl's cousin, the California spotted owl. New regulations included restrictions on clear-cutting, large canopy openings, and prohibitions on removing any trees over thirty inches in diameter. Suddenly, the local timber industry realized that the environmentalists' proposal didn't look so bad after all.

In 1992, three historical adversaries met to discuss a new way forward. They included Plumas County supervisor Bill Coates; chief forester for the largest regional timber company[34] Tom Nelson; and environmental attorney and leader of the Friends of the Plumas Wilderness Michael Jackson. After several initial discussions, they opened up the meetings to other interested residents and moved the venue to the local library (the story goes that one rationale for this location was that it ensured civil interactions and no raised voices). By 1993, the group produced what they called a Community Stability Plan. It called for increased logging in second-growth areas for reasons of local economic development and the reduction of fire risk, including some areas currently protected as reserves. At the same time, the plan protected 148,000 acres of roadless areas located in places slated for logging under current management plans. It also expanded protections for riparian zones and prohibited forty-acre clear-cuts in favor of smaller patch cuts limited to two acres in size. The compromise was intended to offer something to everyone.

Though many community residents were in favor of the plan, opposition from the Forest Service and national environmental NGOs was nothing short of fierce. Agency officials viewed the plan as an undemocratic effort to undermine their decision-making authority as professional foresters committed to the public good. For environmentalists, the plan appeared to weaken national environmental regulations. Besides, the decline in the logging industry was largely of its own making. The effects of harmful past practices combined with growing public support for biodiversity protection on national forests signaled the emergence of new management priorities. Why should environmental groups compromise with the logging industry if the battle was already won?

Implementation of the Community Stability Plan thus required congressional action. Though 140 environmental organizations opposed the bill fearing the precedent it might set to replace national laws with locally defined alternatives, the legislation received heavy bipartisan support, passing the House with a vote 429 to 1. Known as the Herger-Feinstein Quincy Library Group (HFQLG) Forest Recovery Act, since 1998 the plan has been extended twice, in 2003 and again in 2007. Final completion was scheduled for 2012. The Forest Service is currently assessing the effectiveness of the plan for reducing wildfire risk, providing local economic stability, and restoring ecological integrity to the national forests in question.

While Quincy demonstrates the possibilities of collaborative conservation applied to public lands, it also underscores some of the most challenging aspects of implementation. Quincy was not as inclusive of diverse interests or committed to consensus building as some might have wished. For many proponents of collaborative conservation, the fact that the Quincy Group turned to Congress to force through its provisions seemed at odds with the very ideals of collaboration. In hindsight, fears that it would set in motion a string of efforts to weaken environmental statutes have proven unfounded. In this regard, Quincy remains an exception to the rule. A more typical route to collaboration includes the heavy involvement of local Forest Service officials, a greater role for collaborative learning and dialogue, as well as efforts to include a diverse set of local and national interests.

The case of the Ponderosa Pine Forest Partnership (PPFP) illustrates some of these additional dynamics. Centered on rural communities in Southwest Colorado, including Cortez, Mancos, and Durango near the San Juan National Forest (SJNF), the PPFP began about the same time as the Quincy Group and served as a reaction to similar changes in forest management and in society writ large. Here again, collaborative efforts arose within the context of New West socioeconomic changes characterized by the decline of traditional logging, ranching, and mining economies, and the corresponding rise in service-sector industries. The latter trend was accompanied by an influx of new ex-urbanite migrants. The economic tensions experienced by long-term residents still tied to declining industries were soon exacerbated by the emergence of the competing environmental values and land use priorities held by newly arriving residents.

Seeking to avoid the failed confrontational tactics of the County Supremacy Movements taking place in rural counties elsewhere in the West,[35] Montezuma County Commissioner Tom Colbert opted for a more constructive approach. In 1992, he established the Federal Lands Program within the county's administrative offices in order to facilitate better communication and collaborative relationships with officials from the San Juan National Forest, Mesa Verde National Park, and the BLM. Together these three federal agencies accounted for approximately 40 percent of the county's land area.[36] Colbert hoped that by enhancing local-federal relations, more productive solutions to land use problems might present themselves.

In one of the first initiatives of the program, Federal Lands director Mike Preston, SJNF district ranger Mike Znerold, and Colorado Timber Association president Dudley Millard met with Colbert to discuss the situation. Realizing shared interests in forest health and the local economy, they discussed ways to achieve both goals via innovative management ideas. Deciding to focus first on the ponderosa pine vegetation zone on the SJNF (hence, the name), the group secured a small grant from the USDA Rural Community Assistance Program, and the PPFP was born.

The PPFP grew to include some thirty individuals from local, state, and federal agencies, as well as from private industry, academia, private landowners, and local environmental activists. Attempting to show that forest restoration and local economic development need not be mutually exclusive, their plan proposed using local loggers and mill owners to thin national forests in order to reduce the risk of fire, disease, and insect outbreaks. The thinning would be followed by controlled burns in order to reintroduce natural fire ecologies into the landscape. Meanwhile, local loggers would gain access to a steady supply of timber, albeit small-diameter trees, supplemented with a few large logs to enhance the economic viability of the project.

To implement the new plan, the PPFP worked with the SJNF to orchestrate the first-ever noncompetitive timber stewardship contract sold to a local government. Montezuma County agreed to step in as the logging purchaser, which would then subcontract out to local loggers and mill owners. This lessened the financial risk to which small-scale local loggers would be exposed. The PPFP plan paralleled the ideas promoted in Quincy and other small timber towns in the West, and it foreshadowed in many ways the provisions in the Healthy Forest Restoration

Act of 2003. However, the PPFP has had to contend with some daunting and persistent challenges as well.

First, the project suffered early on with the issue of keeping pace with the prescribed fire program. It was one thing to conduct a selective logging operation, but reintroducing fire in a safe manner required very specific weather conditions, as well as sufficient budgetary resources. SJNF foresters soon found themselves falling behind in terms of conducting controlled burns and began to question the long-term feasibility of the program over large acreages. Second, and equally elusive for the PPFP, was the ability of the selective logging operations in small-diameter stands to achieve economic viability. While markets for large-diameter saw logs remained robust, smaller trees still held little value. With the closure of large pulp mills that could use such wood, the PPFP looked to emerging markets in biofuels, furniture, or the development of other innovative products to make the operation profitable for loggers and mill owners. To date, each of these goals remains elusive.

However, like other collaborative conservation efforts across the country, one of the most lasting and profound outcomes of the PPFP was the strengthened relationships among its members—the increased social capital that comes from frequent and productive interaction. The initial effort to conduct forest restoration in the pine zone later expanded to the mixed conifer zone (higher in elevation). And many of the participants in the PPFP became central players in the Forest Service's collaborative effort to revise the comprehensive SJNF Forest Plan. Though both the Quincy Library Group and the PPFP focused on a balance of forest restoration with local economic development, other collaborative groups have emphasized different priorities. The Greater Flagstaff Forest Partnership in Arizona, for example, has stressed the reduction of wildfire risk in the WUI over local economic concerns. And the Catron County Citizens Group in New Mexico has used collaborative conservation to prioritize community cohesion building in the wake of intense local land use conflicts in the mid-1990s.[37] In sum, though groups vary in their structure, participation, and decision-making approach, collaborative conservation appears to be one of the best strategies for redressing and tempering nature-as-commodity thinking by blending timber production with ecological restoration efforts. These efforts, in turn, often seek to remedy the idea of nature as unpeopled, eternal, and pristine by confronting the relatively static vision of nature that has quietly underscored the Forest Service fire-suppression policies over the past century. While these goals have not yet been fully realized, if collaborative processes can help identify ways to improve ecosystem health while providing for the economic needs of local communities in a sustainable manner, then they are certainly worthy of further attention.

The Roadless Area Conservation Rule

Whereas the Quincy Library Group and Ponderosa Pine Forest Partnership reflect grassroots, community-based approaches to national forest management, top-down federal rule changes still play a significant role in shaping forest planning priorities. In addition to national monument declarations, the crafting of the Northwest Forest Plan, efforts to reform mining practices, and the establishment of the BLM's Resource Advisory Councils, one of the most significant public land legacies of the Clinton administration is undoubtedly the Roadless Area Conservation Rule. After completing a three-year study involving six hundred public meetings and the collection of over 1.6 million public comments—more than ever collected for a Forest Service management decision—Secretary of Agriculture Dan Glickman signed off on the Roadless Rule on January 12, 2001. Initiated by Forest Service chief Mike Dombeck, the rule was designed to address the problem of the extensive road system in the national forests that had reached over 343,000 miles by 1985.[38]

For years, the Forest Service viewed road construction as providing a dual public benefit: enabling timber production and providing a resource for motorized and nonmotorized recreation. During the peak of timber production, between 1984 and 1990, the Forest Service oversaw the harvest of some six million acres, producing between 10.5 and 12.7 billion board feet each year.[39] Timber receipts paid to build new logging roads and to restore existing ones. But keeping pace financially with road management was highly problematic. The cost of maintaining these roads to meet proper safety and environmental standards for public users was staggering, and by the late 1990s, the Forest Service carried an $8 billion backlog. By 1997, a mere 38 percent of national forest roads met federal standards. The Roadless Rule allowed the Forest Service to solve this problem by prohibiting new road construction, and therefore most commercial logging, on approximately 58.5 million acres of roadless national forests across thirty-nine states.

But road safety and budget issues were only one part of the rationale for the new rule. Equally important was the shift in management priorities for the Forest Service, in which ecological health and stability increasingly replaced commodity production as the primary measure of successful stewardship. Back in the 1970s, after passage of the Wilderness Act, the Forest Service conducted two inventories of existing roadless areas to assess their potential for wilderness designation (called RARE I and II, respectively; see chapter 8). While some fifteen million acres gained wilderness status in this way, the RARE I and II reviews were criticized harshly by environmental groups for not recommending more land for protection. The Roadless Rule offered another chance to revisit this issue. Although the Rule did not formally designate new wilderness areas, agency proponents were explicit about the benefits the Rule would provide, including wildlife habitat protection, reductions in habitat fragmentation and the number of vectors for invasive species, the protection of watersheds, and the provision of additional recreational opportunities.

To the consternation of environmental organizations, the new Rule did not initially apply to forests covered in the Northwest Forest Plan, nor to the Tongass National Forest in Alaska, which had just completed a new management plan of its own. The Tongass was a particularly sensitive subject. Not only was it the largest national forest in the United States (over seventeen million acres) but also it contained some of the largest remaining expanses of old growth temperate rain forest in the country. However, the Tongass also maintained a long history of federally subsidized timber production, which exacerbated political debates—despite the fact that the local economy was now more reliant upon tourism than logging.[40] Nonetheless, despite these exclusions, the new rule still impacted nearly one-third of all national forestland in the United States.

This environmental victory was short-lived, however. Within weeks of taking office, President George W. Bush suspended the Roadless Rule, citing the need for further review. This action kicked off a political struggle that would continue for the next decade. In May 2001, a federal court challenge brought by the state of Idaho and the Boise Cascade lumber company resulted in a federal judge banning the Rule from taking effect. In response, Earthjustice filed suit, and in December 2002, a federal appeals court reinstated the Roadless Rule. In 2003, new lawsuits were brought against the Rule by the state of Alaska and others. Meanwhile, the Bush administration decided to exempt the Tongass National Forest from compliance, regardless of the court decisions. Then, in 2004, the Bush administration deployed a new strategy: proposing a new Roadless Rule (which the Forest Service adopted in 2005) that allowed state governors to decide if they wished the rule to apply in their own states or not. Efforts by states either to reject or comply with President Bush's new mandate found themselves continually tied up in additional court battles.

Finally, in August 2009, with the Obama administration now in office, the Ninth Circuit Court of Appeals ruled that the original rule would stand in most states (impacting some forty million acres), excluding Idaho and Alaska. However, in 2011, the federal courts changed their position once again, deciding that the rule *would* apply to the Tongass National Forest after all. Finally, in October 2012, the Supreme Court weighed in on the matter, refusing to hear the state of Wyoming's request to review the legality of the Roadless Rule. This decision finally put the matter of legality to rest and allowed its full implementation.

Aside from the marathon legal battles and the tenacity of both sides to keep up the fight for well over a decade, the Roadless Rule case demonstrates the difficulty of pulling away from the dominance of nature-as-commodity thinking in federal land management. Despite the fact that over 90 percent of public comments regarding the rule were favorable and that the status quo was economically unsustainable, opposition to the rule was immediate and extremely potent. The idea that a certain vested interest (in this case the timber industry) would lose access to subsidized lumber was enough to mobilize effective political opposition in the courts, in Congress, and even in the White House. For environmental activists, the case begs the question of whether the Forest Service truly is a multiple-use agency in any meaningful way. Sixty years after passage of the Multiple Use and Sustainable Yield Act, the Forest Service still struggles mightily with the notion that a significant portion of its holdings should be set aside for purposes of noncommercial development. The struggle suggests that the idea of nature-as-commodity remains alive and well in American society, and it continues to wield considerable influence over the ways we use, value, and manage the national forest system.

6

National Wildlife Refuges

In those days, we had never heard of passing up the chance to kill a wolf. . . . We reached the old wolf in time to watch a fierce green fire dying from her eyes. . . . I was young then and full of trigger itch; I thought that because fewer wolves meant more deer, that no wolves would mean hunters' paradise. But after seeing the green fire die, I sensed that neither the wolf nor the mountain agreed with such a view.

—Aldo Leopold[1]

When it comes to wildlife and American society, ours is a complicated relationship. We venerate some species (American bison, for example) and demonize others (wolves). Some stand as symbols of patriotism and national unity (bald eagle), while others divide us more deeply than partisan politics (northern spotted owls, or again, wolves). Certain species can bring warm, fuzzy feelings (river otters), summon tears (passenger pigeons, bison again), or make the hairs stand up on the back of your neck (the grizzly outside your tent, the alligator in the backyard). And still, there are others that inhabit only the shadowy margins of consciousness. We're not really sure what to think about them, until a chance encounter makes us pay sudden and close attention (um . . . is that snake poisonous?).

Of course, the term *wildlife* includes more than just large mammals, reptiles, and raptors. It covers all sorts of flora and fauna, from grasses, trees, and flowers, to insects, fish, songbirds, and everything in between. Nonetheless, wildlife has a way of concentrating and amplifying our complex feelings about nature writ large. In fact, it can do so to a degree rarely matched in disputes over land conservation. Decisions over which wildlife constitute a commercial opportunity, a dangerous threat, a mere nuisance, or a national treasure in need of protection can erupt into emotionally charged conflicts that forcefully mirror the diverse ways we value nature. As the only type of public land specifically designated to protect wildlife, it follows that the national wildlife refuge system evokes equally intense emotional responses.

Or at least one would think if one gave any thought to it at all.

If hard-pressed, many could identify at least one refuge: the Arctic National Wildlife Refuge in Alaska. Perhaps better known by its acronym, ANWR, the refuge is readily identified by those with an interest in environmental politics. The high-profile and passionate debate over whether or not to open ANWR for oil drilling has raged for decades. But beyond this example, how many others can we name?

Currently, the national wildlife refuge system consists of 560 refuges (figure 6.1).[2] Totaling more than 146 million acres, the refuges cover an area larger than the national park system. While there are national wildlife refuges in each of the fifty states, the overwhelming majority of land-based acres—some 85 percent—are located in three huge refuges in Alaska, while another 50 to 150 million acres lie under the sea in the South Pacific.[3]

Of the refuges in the lower forty-eight states, approximately three-quarters are positioned along the north-south flyways of migratory birds. While some refuges emphasize particular species, such as the National Elk Refuge near Yellowstone or the Aransas Refuge in Texas (instrumental in saving whooping cranes from extinction),[4] most target migratory waterfowl and other bird species as part of a broader effort to protect and enhance biodiversity.

With some exceptions, refuges generally prohibit resource extraction activities such as logging, mining, or livestock grazing. However, they all permit a variety of wildlife-related recreational activities, most notably hunting and fishing, which occur in about 300 of the 560 units.[5] While there is an ecologically sound argument for hunting—to replace the role of natural predators in maintaining sustainable population levels—the end result is a seemingly odd situation in which some wildlife species find lower levels of protection in wildlife refuges than in other types of federal lands, such as national parks.

The federal agency charged with managing the wildlife refuge system is the U.S. Fish and Wildlife Service (USFWS). In addition to the refuges, the agency is responsible for an enormous number of related lands and resources, including seventy-one national fish hatcheries, 209 waterfowl production areas (WPAs), fifty coordination areas, forty-six administration sites, and six national monuments. The last category includes two

National Wildlife Refuge System

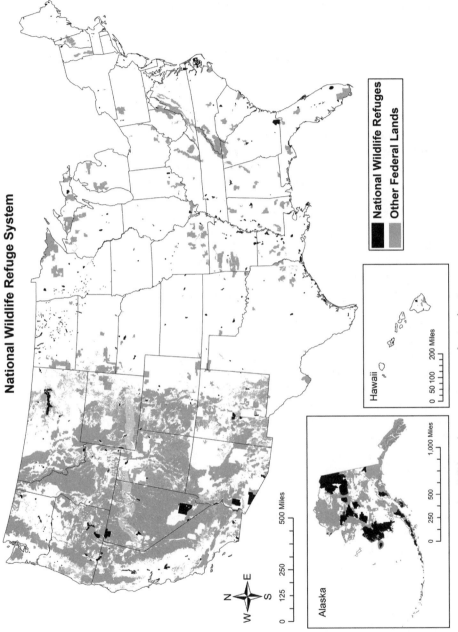

Figure 6.1. The National Wildlife Refuge System. Map by Jessica Lee.

immense marine refuges carved out of the Marianas Trench Marine National Monument in the South Pacific.[6]

The USFWS is best known as one of two federal agencies responsible for implementing the Endangered Species Act (ESA) of 1973.[7] Under this law, the USFWS determines which terrestrial and freshwater species should be listed as "threatened" or "endangered." Significantly, the agency also decides *how* to protect the species in question, a task that includes the identification of critical habitat and the development of a restoration plan. Once designated, this land cannot be used or modified in any way that might impair the recovery of the endangered species in question, unless government approval is granted. As one might imagine, this can be terribly controversial, especially when the necessary habitat spills over onto privately owned lands.[8]

However, one might also imagine that the primary rationale for establishing wildlife refuges is to accommodate endangered species listed under the ESA. But interestingly, only 59 of the total 556 wildlife refuges were created primarily for this purpose.[9] Nonetheless, though the number keeps changing, approximately 280 of the over 1,200 species listed as endangered or threatened under the ESA are currently found on refuge lands. And one can argue that the degree to which the national wildlife refuges have *kept* species from being listed as endangered is also a mark of the system's success.

But the questions don't end there.

As mentioned earlier, it might be reasonable to expect that wildlife is afforded greater protection in a national wildlife refuge than in other types of public lands, but this is not always the case. Similarly, one might assume that the public organizations advocating for the national refuge system are fully committed to environmental preservation rather than resource use. But as we shall see, if one considers the role of sport hunting groups in cultivating public support for the system, this assertion also comes into question.

Then there is the matter of public awareness. With its vast acreage, long history, and designated federal management agency, one might imagine that the national wildlife refuge system lies at the center of conservation efforts in the United States. But instead, it has struggled to garner public attention compared to other types of public lands.[10] Given the passion that many people express for wildlife, why have the refuges not played a larger role in our national narrative of environmental protection? Geography—with the system's heavy Alaskan (and southern Pacific) focus—may provide one answer, but it is hardly satisfactory, since it may be as much effect as cause.

Hence, the question remains: Why has it been "easier" to achieve decisive success in efforts to set aside a grand canyon in Arizona or a geyser in Wyoming than to protect northern spotted owls, grey wolves, or Pacific salmon?[11] Perhaps the bigger mystery is why a conceptual separation even exists at all between landscapes and the wildlife that live there. The national wildlife refuge system embodies each of these riddles and apparent contradictions. Making sense of them takes us back once again to consider the legacy of ideas underlying the broader environmental movement: the continuity of commodity thinking and conceptions of nature as unpeopled, eternal, and pristine. The constant bumping up of these diverse values with efforts to protect and preserve nature creates tensions that persistently challenge the wildlife refuge system. In this sense, the refuges are perhaps more akin to ideological battlegrounds than places of respite. And they have been this way since the very beginning.

WHO OWNS WILDLIFE? STATE RIGHTS AND THE SEPARATION OF LAND AND LIFE

As noted in earlier chapters, viewing nature as a commodity served as a basic tenet of the colonial enterprise in North America. Wildlife held particular economic value as a provider of furs, leather goods, food, and even high-fashion accessories. As such, wildlife resources were of keen interest to the states after independence. But unlike the public domain, there is no specific provision in the Constitution authorizing federal authority over wildlife. Into this void, the individual states asserted their own jurisdiction, setting up their own game laws and institutions for enforcement.

However, state authority (and even that of the federal government) was premised on the insistence that wildlife remained outside the purview of private property. When one purchased land, it did not follow that one also gained ownership of the wildlife found there. Instead, it remained in the public trust. This so-called Public Trust doctrine[12] was confirmed in two notable Supreme Court decisions, in 1842 and 1896, respectively.

In the first case, *Mater v. Waddell*, a riparian landowner claimed ownership not only of his property and riverbanks but also the portion of the river flowing through it, including the oyster beds. For proof, he provided a title dating back to a 1644 transfer from King Charles of England to the duke of York. The Supreme Court, however, did not agree. Referring back to English Common Law, the Court ruled that originally, the "navigable waters and lands under them" were held in public trust by the king.[13] After U.S. independence, that trust now passed to the state of New Jersey, not the individual property owner.

An 1896 case reinforced this decision. In *Greer v. Connecticut*, the defendant had tried to move lawfully harvested game birds out of Connecticut, which was against state law. Ignoring the issue of interstate commerce, which may have led to questions over federal authority in the matter, the Supreme Court ruled against Greer on the basis that the "right to control and regulate the common property in game" resided with the states. Moreover, this right was to be carried out "as a trust for the benefit of the people."[14]

These two court decisions, along with several others, set the legal foundations for state authority over wildlife. Throughout the nineteenth century, state governments moved to pass game laws establishing hunting seasons and setting limits on the number of fish, birds, or other wildlife hunters could take. Massachusetts and New Hampshire were the first to establish game wardens in 1850, while New Jersey and Connecticut led the way in protecting nongame bird species that same year.

All of this helps to explain the fragmented management situation that exists today on many U.S. public lands. For example, the U.S. Forest Service manages the land, water, trees, and other vegetation on a national forest. But legally, the agency holds no jurisdiction over the wildlife that resides there. Instead it is managed by a *state* department of wildlife. So here we have it: two different agencies from two different levels of government, attempting to manage two parts of the same ecosystem . . . habitat and wildlife, land and life. If one adds to this mix a divergent set of management priorities, whereby wildlife agencies are faced with a powerful incentive to promulgate game species, sometimes at the expense of nongame species, in order to ramp up sales of hunting and fishing permits, the picture gets even more convoluted. Surely such jurisdictional fracturing can only hamper efforts to manage public lands in a holistic way.

But the story does not end here. If the states had matters under control, one might ask, why did the federal government get involved in the business of wildlife management in the first place? Why do we need a *national* wildlife refuge system, or, for that matter, a U.S. Fish and Wildlife Service to oversee it?

SPORT HUNTING AND CONSERVATION, OR WHEN A REFUGE IS NOT A REFUGE

As detailed in chapter 3, nineteenth-century settlement and industrial expansion led to the decline or outright eradication of many species in the United States. Threats to wildlife came in many forms. Originally, subsistence and commercial hunting had the greatest impact. The emergence of European markets for items such as beaver pelts, bison robes, and leather goods created enormous incentives for the decimation of various species. Settlers' efforts to clear land for cultivation and use as pasture resulted not only in the loss of wildlife habitat, but the importation of non-native diseases carried via domesticated livestock. These developments were often followed by rigorous attempts to extirpate predators. Finally, in the post–Civil War era, the development of sport hunting added a new source of pressure on wildlife populations.

Needless to say, early regulatory efforts by state governments were largely ineffective. In fact, for migratory species such as waterfowl, state game wardens and commissioners often strove to set bag limits higher than neighboring states in order to allow their citizens a greater share of transient wildlife resources—a classic tragedy of the commons if ever there was one. Clearly, there was a need for federal intervention to rein in the destruction.

The impetus for change, however, arose from a singularly unlikely source. One of the great ironies in American conservation history is that the most strident efforts to protect wildlife were orchestrated not by Romantic preservationists, but by those who most loved to shoot it (and I do not mean with a camera). While preservationists such as George Catlin, Henry David Thoreau, and John Muir are famous for making early calls to protect wildlife, the actual mobilization of political and economic power to achieve these ends emanated from the sport hunting community. As early as 1844, elites in New York organized hunting clubs that advocated for protected game lands.[15] With members drawn from wealthy and influential circles of nineteenth-century American society, the clubs purchased land and organized expeditions. Armed not only with rifles, but also with law degrees, members frequently pursued poachers and game law violators on their own. State laws allowed for citizen suits against such individuals, and the clubs were tremendously effective in this regard. Sport hunting organizations soon sprouted up in larger cities around the country.

If one sees parallels here between these nineteenth-century game parks and the private hunting grounds of European royalty, the comparison is not far off the mark. The elites of the sport hunting community saw hunting as a recreational exercise, rather than something conducted for subsistence or commercial purposes. These early conservation efforts, therefore, carry heavy class overtones. Key to their efforts was the distinction between sport or leisure hunting and so-called market or commercial hunting. Market hunters might be small scale, simply attempting to put dinner on the table or to augment their incomes by selling a few animals each week. Or they might be professionals working on a much larger scale, capable of killing thousands of animals in a single afternoon for profit (as was often the case with bison). Of course, sport hunters were also capable of such wasteful slaughter if they did not practice the values of wildlife preservation espoused by many hunting clubs.

One of the most outspoken advocates of these values was George Bird Grinnell. Born to a wealthy family in New York, Grinnell grew up near the estate of Lucy Audubon, widow of famed naturalist John James Audubon. From Lucy he acquired a deep appreciation for nature and a love for the outdoors. As both a hunter and the publisher of *Field and Stream* magazine, Grinnell used his influence to call for the creation of sport hunting associations and to denounce any industry that promoted the destruction of wildlife. A favorite early target was the fashion industry, which at the time advocated the use of colorful bird plumage (and other body parts) as desirable accessories for women's hats. But he also brought hunters to task when he felt they failed to adhere to good practice. When readers sent him photos of their numerous hunting trophies expecting praise, he might criticize them publicly in the pages of his magazine, labeling them "game hogs" if they took more than their fair share.

In 1886, Grinnell teamed up with Justice Oliver Wendell Holmes Jr., writer John Greenleaf Whittier, and others to establish the Audubon Society. Named after his early mentor, the group focused on the protection of wild birds and their eggs. Each member signed a pledge "not to molest birds," and after the first year, the group totaled over thirty-nine thousand members. New chapters soon began springing up across the country.

Still, Grinnell remained an avid sports hunter. The following year, in 1887, he worked with Theodore Roosevelt to create the nation's most famous sport hunting association, the Boone and Crockett Club. The group advocated for responsible big game hunting, including protections for animals such as the American bison. One of the group's first campaigns was to protect the large game in Yellowstone National Park. The park's foundational 1872 legislation did not explicitly protect wildlife from hunting. Years earlier, Grinnell, General Philip Sheridan, and Senator George Vest of Missouri had tried and failed to pass congressional legislation to expand Yellowstone and strengthen hunting regulations. While the U.S. Cavalry now patrolled Yellowstone National Park as de facto managers (thanks to General Sheridan), poaching laws remained weak.

Refusing to give up on the issue, Grinnell decided in 1893 to include a feature story in *Field and Stream* about the poaching problem in Yellowstone. He assigned reporter Emerson Hough to the task. Shortly after arrival in the park, Hough was fortuitous enough to encounter one of Yellowstone's most infamous poachers. The man was caught in the act of skinning one of five dead bison, the bodies still warm from recent shooting. Armed now with photographs of the carnage and boastful remarks from the poacher himself, Hough crafted a powerful story that swayed public opinion on the issue. The following year, Congress passed the Yellowstone Game Protection Act of 1894.

Sponsored by Representative John F. Lacey from Iowa, one of the most stalwart supporters of the early conservation movement, the Act made it unlawful to hunt wildlife in the park or remove it from park boundaries. It also banned any activities that might threaten wildlife and gave power of enforcement to the Interior Department. Twenty-two years after the creation of Yellowstone, the nation finally had its first federal wildlife protection law. And at the same time, it created what was in effect, if not in name, the nation's very first wildlife preserve: Yellowstone National Park.

When is a refuge not a refuge? When it is a national park.

Though for preservationists and nonhunters, there is another way to answer this question; namely, when the so-called refuge is simultaneously a public hunting ground. But that story comes a bit later.

THE FIRST (ACTUAL) NATIONAL WILDLIFE REFUGE

Creating the first federal wildlife refuge that was *not* a national park took a few more years.[16] The next step in this process occurred in 1900 with passage of yet another law sponsored by Representative Lacey. Bearing the representative's name, the Lacey Act carefully revisited and expanded the Supreme Court's 1896 *Greer* decision to allow for federal intervention in wildlife matters. It did so by interpreting the transportation of wild game across state lines as an issue of interstate commerce, and therefore, subject to federal regulatory authority under the Constitution. In this way, the Lacey Act made it a federal crime to transport any wildlife killed in violation of state laws across state lines. The act also allowed states to prohibit the transfer of *legally* killed game across state boundaries. Finally, it gave the Secretary of Agriculture power to restore, introduce, distribute, and preserve game birds in accordance with state laws, a mandate with implications for regulating the movement of non-native and invasive species.

While the Lacey Act strengthened the hand of the federal government regarding wildlife conservation, there still existed no designated unit within the public land system specifically designed for wildlife protection. This changed in 1903 thanks, once again, to President Teddy Roosevelt. Given Roosevelt's support for national parks, forests, and other conservation measures, members of the Florida Audubon Society and American Ornithologists' Union decided in 1901 to request his assistance in a long-running campaign to protect Pelican Island in Florida. The 5.5-acre mangrove island was the last remaining rookery (or nesting colony) on the East Coast for brown pelicans and over sixteen other bird species. One of these species, the egret, was under intense pressure from the fashion industry, which used the bird's plumage to decorate women's hats. Though state protection against hunting on Pelican Island was achieved in 1901, regulations were largely ignored. In fact, two of the first four wardens hired to protect the island were murdered in the line of duty. Yet because the island was part of the public domain, President Roosevelt wondered if federal intervention might be possible. He famously asked his aides if there was any law preventing him from declaring it a sanctuary for birds and other wildlife. Hearing of none, on March 14, 1903, he issued an executive order naming Pelican Island the nation's first federal bird reservation—the first unit of what would later become the national wildlife refuge system.[17]

The president's order directed the Department of Agriculture's Biological Survey Office to manage the reservation. However, like the first national parks, no federal funding was allocated for management purposes. To fill the gap, members of the Audubon Society volunteered to serve as custodians, all the while working to identify sites for additional federal bird reservations across the nation. In 1906, their efforts received a boost with passage of the Antiquities Act, which gave the president additional authority to carve out new national monuments from the public domain. By the time Roosevelt left office in 1909, he created fifty-one bird reserves, many of which later became wildlife refuges. But of course, as a founding member of the Boone and Crockett Club, he was equally interested in protecting big game species as well.

Back in 1901, President McKinley had created the Wichita Forest Reserve in Oklahoma on lands deemed unsuitable for farming. Three years later, the Boone and Crockett Club and the New York Zoological Society launched a campaign to save the remaining American bison by creating a refuge on fifty-nine thousand acres of grassland within the Wichita Reserve. In 1905, President Roosevelt designated Wichita as a game reserve, and two years later, he transferred fifteen bison from the New York Zoo to begin reestablishing the species on the southern Plains. The success in Wichita triggered the formation of the American Bison Society, which lobbied Congress in 1908 to establish the National Bison Range in the northern Plains, on the Flathead Indian Reservation in Montana. With bison herds now reestablished in Yellowstone, the Wichita Refuge, and the National Bison Range, the species was accorded a new lease on life. In 1912, elk received a similarly dedicated rangeland with the creation of the National Elk Refuge near Yellowstone National Park and the town of Jackson Hole, Wyoming.

GOING INTERNATIONAL TO SAVE THE NATIONAL COMMONS

Despite gains made during the Roosevelt administration, by the mid-1910s, state and federal efforts to regulate the hunting of wildlife were again falling short. Roosevelt's new reserves and passage of the Lacey Act helped to slow the trade of wild game, but many migratory species continued to experience a sharp decline. Some states still allowed spring hunts of migratory birds and encouraged hunters to take as many as possible before they flew on to neighboring states. Bag limits remained weakly enforced or were set too high to be sustainable.

In response, Senator Elihu Root of New York sponsored a resolution authorizing the president to enter into international treaties to protect migratory species. In 1916, President Wilson did just that, brokering a treaty between the United States and Great Britain (acting on behalf of Canada). Two years later, Congress passed the Migratory Bird Treaty Act of 1918, giving the treaty force of law. The agreement protected some eight hundred species of migratory birds that regularly traveled between the United States and Canada. It became a federal crime to "pursue, hunt, take, capture, kill, attempt to take, capture or kill, possess, or offer for sale" certain species of birds, including their nests and eggs.

Arguing that the treaty undercut their authority over wildlife, several states brought a legal challenge in the Supreme Court. In *Missouri v. Holland*, Chief Justice Oliver Wendell Holmes declared:

> But for the treaty and statute, there soon might be no birds for any powers to deal with. We see nothing in the Constitution that compels the government to sit by while a food supply is being cut off and the protectors of our forests and crops are destroyed. It is not sufficient to rely upon the States.[18]

With the jurisdictional matter thus settled, the Bureau of Biological Survey was assigned the task of setting national regulations pertaining to the named migratory species. Additional agreements followed between the United States and Mexico, the Soviet Union (Russia), and Japan to protect other migrating species.

By the 1920s, the populations of many migratory bird species rebounded dramatically. However, this growth coincided with a new problem: the deterioration of habitat. Though overhunting was now less of a concern, the growing populations of migrating birds now faced a shortage of places to rest on their long North–South migrations. In both the Midwest and eastern United States, this was especially problematic as wetlands were drained for agricultural production or urban growth. Once again, sport hunting clubs and conservation organizations mobilized for action. Since these eastern flyways corresponded with privately owned lands, it was necessary to lobby Congress for money to buy the needed wetlands. In 1924, Congress established the Upper Mississippi Wildlife and Fish Refuge, a 284-mile-long ribbon of land covering 194,000 acres across four states along the banks of the Mississippi

River. To accommodate hunting groups, sport hunting was allowed on the refuge, creating the nation's first "public shooting ground." This compromise was not enthusiastically received by antihunting advocates, including some Audubon members, but it was accepted as a necessary evil to gain congressional support.

Five years later, recognizing the need for additional refuges along the Midwestern and eastern flyways, Congress passed the Migratory Bird Act of 1929. This law created a commission consisting of congressional representatives and Cabinet members to identify and oversee land acquisitions for new wildlife refuges. It also allocated funds for this task. However, with the onset of the Great Depression, implementation quickly ground to a halt. Movement on wildlife refuge issues did not resume until 1934, when the Franklin D. Roosevelt administration established a new committee on wildlife. With membership that included Aldo Leopold, publisher Thomas Beck, and famous *Des Moines Register* cartoonist J. N. "Ding" Darling, the committee recommended the federal purchase of some $50 million in new refuges. To cover costs, they proposed that hunters buy a "duck stamp" each year, with proceeds going to the acquisition of new refuge lands identified and authorized under the 1929 Migratory Bird Act. Later that year, the Duck Stamp Act passed, cementing what would be a long-term mutual association between the sport hunting community and national wildlife refuges.[19]

FDR's next move was to name Darling as the new head of the Bureau of Biological Survey. As a longtime critic of the New Deal for not going far enough in its conservation efforts, Darling's appointment came as something of a surprise. But his boundless energy and adeptness at acquiring federal funds allowed the new Bureau director to expand the refuge system considerably. Vital to his success was Darling's right-hand man, biologist J. Clark Salyer II. Driving thousands of miles across the country in his worn-out Buick, Clark identified, assessed, and cataloged over six hundred thousand acres of potential lands for purchase, leading to the creation of over fifty-five new refuges. These included Red Rock Lakes Refuge in Montana that protected trumpeter swans, and the Agassiz Refuge in Minnesota that preserved habitat for moose.[20] One of Darling's crowning achievements came in 1936, when he worked with FDR to convene the first North American Wildlife Conference in Washington, D.C. Bringing together over two thousand sport hunters, anglers, ornithologists, garden club members, 4-H enthusiasts, and others interested in supporting wildlife, Darling used the occasion to launch the National Wildlife Federation (NWF), the nation's first and largest organization explicitly devoted to the conservation of all wildlife species.[21]

BUILDING A FEDERAL WILDLIFE AGENCY

The growth of federal laws, international treaties, national NGOs, and the expansion of the national refuge system in the 1930s did not spell the end of state management activity. In fact, in 1937, the Pittman-Robertson Act provided federal monies to the states to create their own refuge system. While conservationists generally applauded this expansion, it also perpetu-ated the fragmented dual-government approach to wildlife management.

Meanwhile, at the federal level, the nation's wildlife refuges and game preserves still lacked a dedicated management agency. Originally, the Bureau of Biological Survey had been tapped to fill this role, but it did not have significant budgetary or personnel resources for the task. This changed with passage of the Reorganization Acts of 1939 and 1940. Bringing together the Bureau of Fisheries from the Department of Com-merce and the Bureau of the Biological Survey from the Department of Agriculture, a new agency was created: the Fish and Wildlife Service (FWS) in the Department of the Interior. Under the leadership of Ira N. Gabrielson, the first director, the FWS was granted managerial authority over the nation's wildlife refuges.

This sounds simple enough and perhaps reasonably explains the strange title of the Fish and Wildlife Service (fish are wildlife too, no?). However, one need only look more closely at the bureaus and depart-ments listed above to notice something strange. What was the Bureau of Fisheries doing in the Department of Commerce anyway? And what in-terest did the Agriculture Department have in biological surveys of wild animals? Surveys of domesticated species made sense—cattle, sheep, hogs—but wildlife by definition seemed out of place on the farm.

The Bureau of Fisheries dates back to an 1871 congressional resolu-tion authorizing the president to create the U.S. Fish Commission in the Department of the Treasury. The purpose was to oversee freshwater and marine fishing activity, which a special focus on the Alaskan salmon industry. In 1903, the Commission was merged with a newly created Bureau of Fisheries in the Department of Commerce and given the dual— and contradictory—mandate to both *protect* and *promote* the development of fish resources in the United States. Not unlike the contradictory charge given to the National Park Service, these two opposing goals proved untenable for the Bureau. Given the lack of budgetary resources and po-litical support, by all historical accounts, the Bureau was profoundly inef-fective in terms of carrying out its regulatory duties and remained unduly influenced by the industry it was supposed to police.[22]

The other half of the new Fish and Wildlife Service (e.g., the Bureau of Biological Survey) shared a similarly checkered past. Created in 1886, the

Biological Survey evolved from what had been established only one year before as the Division of Economic Ornithology and Mammalogy. Early tasks consisted of mapping and identifying the distribution of plants and animals in the United States. The initial purpose of this inventory was to ascertain which species might prove economically useful to agriculture.[23] This explains the presence of the Biological Survey within the Agriculture Department. However, as the Bureau evolved, it gained additional responsibilities, including the oversight of Teddy Roosevelt's wildlife preserves and, increasingly, the nation's predator control programs.

As early as 1905, the Bureau developed protocols and instructional materials for dealing with predator/livestock problems on western ranches. In 1913, to address a plague outbreak on national forests, the Biological Survey initiated rodent control operations. However, a turning point was reached in 1916. That year the agency received more congressional funding for predator control than for any other activity, including its biological inventory.[24] Supplanting the Forest Service's predator control program, the Biological Survey engaged heavily in a number of eradication initiatives, including the elimination of coyotes due to a rabies outbreak in 1916 and a rabbit population problem that surfaced in 1920. Later, it took on the task of protecting the nation's grain supply from the threat of rats. Methods of control included trapping, shooting, and increasingly, the use of poisons or "toxicants." In 1931, Congress passed the National Animal Damage Control Act to ensure future funding for predator control efforts, including the development of more effective toxicants.[25]

Even after the Biological Survey transformed into the new Fish and Wildlife Service in 1940, predator control continued to be a major responsibility. Although the development and use of poisons was halted in 1971 after lawsuits brought by the Defenders of Wildlife, the Sierra Club, and the Humane Society spurred President Nixon to issue Executive Order 11643, predator control by other means continued until 1986. That year Congress finally moved the Animal Damage Control Program from the Fish and Wildlife Service back to the Agriculture Department (changing the name to Wildlife Services in 1997). Nonetheless, for its first forty-seven years, the same federal agency charged with protecting and managing wildlife in the United States was also responsible for the control and, at times, eradication of certain species.[26]

In sum, the bureaucratic genealogy of the new Fish and Wildlife Service was complicated to say the least. One half (evolved from the Bureau of Fisheries) historically viewed wildlife as a commodity and maintained close ties to industrial interests. Meanwhile, the other half (evolved from the Biological Survey) appeared to be equally committed to wildlife "control" as it was to conservation. Meanwhile, the most politically powerful constituency of the Fish and Wildlife Service continued to be the sport hunting community.

Perhaps unsurprisingly, the task of clarifying conservation as the primary mission of the agency was still several decades away.

BUT WHAT ARE REFUGES FOR?

Despite creation of the new federal agency, the influence of hunting groups on the national wildlife refuge system did not readily diminish. In fact, in 1949, their power appeared to be on the rise when FWS officials lobbied Congress to pass an amendment to the Duck Stamp Act. The new law doubled the cost of a stamp from $1 to $2, but at the same time, expanded the number of wildlife refuges open to hunting. It did so by redefining one-quarter of existing "inviolate sanctuary refuges" (where hunting was prohibited) into "wildlife management areas," units in which hunting was allowed. While this action doubled the revenue available for managing and expanding the refuges, it came at a steep cost. For preservationists and other wildlife advocates outside of the sport hunting community, the FWS and refuge system were in dire need of institutional reform.

However, when change did come, it seemed to be moving in the wrong direction. In 1956, the agency received its first reorganization via the federal Fish and Wildlife Act. This law split the agency into two divisions, a Bureau of Sport Fisheries and Wildlife (given charge of the nation's wildlife refuges) and a Bureau of Commercial Fisheries, all under the umbrella organization of what was now called the *United States* Fish and Wildlife Service (hence today's acronym, USFWS). While this action stressed the agency's commitment to hunting and angling along with the commercial development of the U.S. fishing industry, conservation received nary a mention.[27]

Then, in 1958, Congress once again amended the Duck Stamp Act. In exchange for an increase to $3 per stamp, the USFWS agreed to open up 40 percent of remaining inviolate refuges to sports hunters. The law also included authorization for a Waterfowl Production Area (WPA) program to buy and protect vital migratory bird wetlands in the Prairie Pothole Region of the Midwest. Significantly, the WPAs would not be subject to the 40 percent limit on hunting, but completely open to such activities. In 1961, the WPA program received a boost when the Wetlands Loan Act authorized a $200 million loan to the USFWS (to be paid back with Duck Stamp receipts) to accelerate the purchase of threatened wetlands over the next twenty-three years. The following year, Congress passed the Refuge Recreation Act of 1962, authorizing the USFWS to open refuges to a wide variety of recreational uses as long as they did not conflict with their primary purpose.

In 1966, the National Wildlife Refuge System Administration Act formally established the National Wildlife Refuge System in the United States. The various federal bird, fish, and game preserves, WPAs, and fish hatcheries were brought together under the single regulatory roof of the USFWS. However, despite its monumental title, the 1966 act failed to provide a clear mission statement (or "organic act") for the agency or system. Instead it offered only broad guidance for refuge management and reconfirmed that all recreational activities and uses be "compatible" with the refuges' stated purpose.

For the first twenty-six years of its existence, the Fish and Wildlife Service oversaw a refuge system increasingly defined as a resource for sport hunting, fishing, and recreational activity. This is not to say that many local FWS managers did not do their best to responsibly manage and protect the resources placed in their trust. But according to Zaslowsky and Watkins, higher levels of management within the agency clearly embraced recreational priorities.[28]

Returning to a question posed earlier: when is a refuge not a refuge? For antihunting preservationists at least, it is when the Bureau of *Sport* Fisheries and Wildlife managed them as public shooting grounds, fishing ponds, and play areas for motorized watercraft. While acknowledging the need for population control via hunting, wildlife advocates nonetheless continued to pose a different question: Where was the conservation mandate?

TURNING THE CORNER TO CONSERVATION

Environmental historians and political scientists often point to Rachel Carson's *Silent Spring* as one of the primary catalysts for the modern environmental movement in the United States. Yet fewer may be aware that she actually began her career as a marine biologist with the Bureau of Fisheries, and later, the U.S. Fish and Wildlife Service. In 1947 she served as editor for the agency's *Conservation in Action* series, a set of publications designed to spur public interest in and appreciation for wildlife conservation efforts. In 1951 she published *The Sea Around Us*, which explained the ecology of marine environments to a general audience. Due to its success, she retired from the FWS to write *Silent Spring*. Published in 1962, the book offered a powerful critique of the damage caused by pesticides to both humans and wildlife.[29]

However, despite her professional association with the agency, it is less clear what impact Carson had on moving the FWS toward a more conservation-oriented set of priorities during her years as an employee. Nonetheless, the growth of national concern for environmental protection during the 1960s and 1970s undoubtedly rendered profound and lasting effects. New responsibilities, management mandates, and a rapid expansion of the system stretched the Fish and Wildlife Service to the limit. By the time the dust settled, the agency was firmly committed to a new set of conservation priorities for the national wildlife refuge system as a whole. One of the first new legislative mandates came from the 1964 Wilderness Act. This law ordered all federal agencies to inventory their lands for potential designation as wilderness areas. In 1968, the Great Swamp National Wildlife Refuge in New Jersey became the first refuge to have a wilderness area administered by the USFWS.[30] Today, the agency is responsible for over twenty million acres of wilderness areas located within the national wildlife refuge system.

Another new mandate arrived in 1969, when Congress passed the National Environmental Policy Act. The law required all federal agencies to seek consultation on potential environmental impacts, including those upon wildlife, for all management projects. The USFWS soon found itself scrambling to fulfill this new mandate, serving as wildlife consultant for all other federal agencies in addition to its regular duties.

In 1970, the Bureau of Fisheries was pulled from the USFWS and returned once again to the Commerce Department. Renamed the National Marine Fisheries Service, it became part of the newly created National Oceanographic and Atmospheric Administration. The next year, a much leaner and focused USFWS became authorized under the Alaska Native Claims Settlement Act of 1971 (ANCSA) to transfer enormous tracts of wildlife habitat into the national refuge system. The USFWS set to work, assessing some eighty million acres of Alaskan land for potential designation as federal wildlife refuges.

And then things got really busy.

In 1973, passage of the Endangered Species Act resulted in an exponential increase in the USFWS's regulatory duties. Named in the law as the agency responsible for determining threatened or endangered listings for all terrestrial and freshwater species, the USFWS's workload exploded. But listing was only one aspect of this new mandate. In addition, the USFWS acquired the task of defining critical habitat and developing restoration plans for each species. Suddenly, this "behind the scenes" agency was attracting massive public attention. The politically and economically charged task of implementing the ESA necessarily placed the FWS in the national spotlight, and not always favorably.

The next year, the agency received a rather confusing name change. Recall that since 1956, the Bureau of Sport Fisheries and Wildlife was the agency officially in charge of wildlife refuges (serving under a shadow umbrella bureaucracy titled the USFWS). Dropping the word *sport* and eliminating the umbrella organization, the agency was now simply called the U.S. Fish and Wildlife Service, a name that no longer privileged recreation, but reflected a stronger commitment to wildlife conservation (figure 6.2).

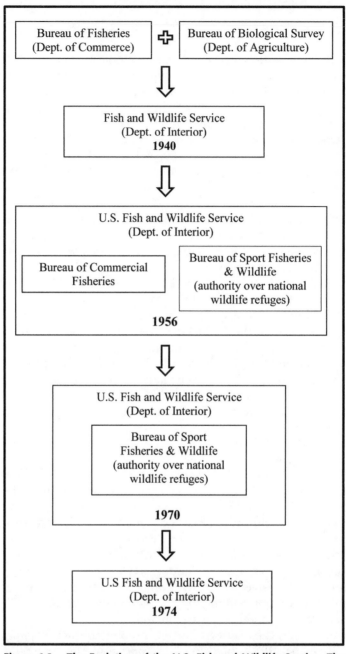

Figure 6.2. The Evolution of the U.S. Fish and Wildlife Service. The Fish and Wildlife Service was created in 1940 with the merger of the Bureau of Fisheries and Bureau of Biological Survey. In 1956, two bureaus were created *within* the Fish and Wildlife Service: the Bureau of Commercial Fisheries and the Bureau of Sport Fisheries and Wildlife, with the latter given responsibility for national wildlife refuges. In 1970, the Bureau of Commercial Fisheries was removed. Finally, in 1974, the Bureau of Sport Fisheries and Wildlife merged back into the umbrella U.S. Fish and Wildlife Service to form a single administrative entity.

In 1976, Congress added to the agency's conservation duties by moving "game ranges," previously managed jointly with the Bureau of Land Management, entirely to the USFWS. And finally, in 1980, the Alaska National Interest Lands Conservation Act (ANILCA) expanded seven existing wildlife refuges and created nine more. This move was truly monumental. By establishing over fifty-three million acres of new national wildlife refuges, ANILCA nearly tripled the size of the system overnight. Alaskan refuges now accounted for some 76.4 million acres, or 85 percent of the entire system.

After passage of ANILCA in 1980, the system of national wildlife refuges continued to expand, albeit at a slower rate. While Congress contributed a handful of new refuges, the primary mechanism for expansion tended to be presidential declarations via executive order. Every U.S. president since Teddy Roosevelt established new wildlife refuges in this way, and the pattern continued with President Reagan, who created thirty-four new refuges during his two terms. President George H. W. Bush then added forty-one refuges, while President Clinton declared fifty-six more.[31] Since the Clinton administration, however, the pace has slowed considerably. President George W. Bush established only twelve in eight years (though measured in terms of acreage, his contributions are of historic proportions, see below). And in President Obama's first term, the administration oversaw the creation of four wildlife refuges, four conservation areas, and one wildlife management area.[32]

On the regulatory front, however, presidential administrations have oscillated wildly in their support, largely in keeping with partisan ideologies. Both the Reagan and G. H. W. Bush administrations clashed repeatedly with Congress amid efforts to reduce USFWS budgets and to redefine managerial priorities to better promote resource development. A true low point for the USFWS came during the long-running and highly publicized conflict over the northern spotted owl in the Pacific Northwest. Charged with politicizing what was supposed to be a science-based listing decision under the Endangered Species Act, the USFWS was forced by a federal court to reverse an earlier decision and declare the owl a threatened species. At the same time, the agency faced constant administrative pressure to open wildlife refuges, including the Arctic National Wildlife Refuge, to increased oil and gas development and other activities viewed by environmentalists as detrimental to wildlife (see below). Matters came to a head in 1993, when the USFWS and Department of the Interior were forced to settle a federal lawsuit brought by the Audubon Society, Wilderness Society, and Defenders of Wildlife to terminate specified activities on wildlife refuges deemed not compatible with their purposes.[33]

Despite these challenges, some significant progress was made in 1989 with passage of the North American Wetlands Conservation Act, a law that implemented a 1986 treaty with Canada. Specifically, the act provided federal matching funds to states, NGOs, and private landowners to purchase, donate, or establish conservation easements on valuable wetland habitat used by migratory birds. Congress also passed a number of laws implementing fisheries protection treaties for species including salmon, halibut, and tuna.

However, the tide really turned in 1997. That year, following an executive order signed by President Clinton, Congress passed the National Wildlife Refuge System Improvement Act. Finally, the wildlife refuge system received its long-awaited "organic act." Amending the 1966 Act, the new law explicitly identified wildlife conservation and the maintenance of biological integrity as the primary missions of the system. In addition, it defined and prioritized appropriate wildlife-dependent recreational activities, including hunting, fishing, wildlife observation, photography, interpretation, and environmental education. Finally, the law mandated long-term comprehensive management plans and codified a process for determining refuge use compatibility.

During the Clinton years, the refuge system experienced several significant expansions, including massive new marine sanctuaries in the South Pacific near the U.S. territories of Guam, Palmyra Atoll, Kingman Reef, and the Midway Atoll. In addition, some fifty-seven thousand acres of salmon habitat gained federal protection on the Columbia River. Efforts to address the anonymity of the wildlife refuge system received a boost with the 1998 passage of the National Wildlife Refuge System Volunteer and Community Partnership Enhancement Act. This law encouraged public participation in USFWS programs designed to build awareness and education regarding the national refuge system. The following year, the USFWS adopted Puddles the Blue Goose as the official agency mascot. Standing proud beside other public land icons such as Smokey Bear and Woodsy Owl, Puddles sought to put a recognizable and welcoming face to the USFWS for kids, their parents, and the general public (figures 6.3, 6.4, and textbox 6.1)

Figure 6.3.   Logos for the U.S. Fish and Wildlife Service (left) and the National Wildlife Refuge System (right). The latter features a stylized image of Puddles the Blue Goose. Image courtesy of the USFWS.

Figure 6.4. Puddles the Blue Goose came in first place in the Twin Cities Mascot Marathon held in October 2012. According to the U.S. Fish and Wildlife Service press release, Puddles "skipped the fall migration to train for the race." Photo credit: Beth Ullenberg, courtesy of the USFWS.

TEXTBOX 6.1.
The Story of Puddles

"Puddles who?" you ask. If you are a bit befuddled by the large, fuzzy blue goose, you are not alone. Though relatively unknown to the general public, and certainly not in the same league as Smokey Bear or Woodsy Owl, Puddles the Blue Goose is the charismatic mascot of the U.S. Fish and Wildlife Service. And his star is on the rise!

In fact, Puddles has been around for decades, though not necessarily known by that name. Back in the 1930s, cartoonist and first FWS director J. N. Ding Darling sketched a number of animals that were featured on various national wildlife refuge signs. One of the most popular was the image of a flying blue goose. Even Rachel Carson wrote about the ubiquitous waterfowl. In an essay introducing the FWS Conservation in Action book series, a set of narratives about the Refuge System, she noted the following:

> As you travel much in the wilder sections of our country, sooner or later you are likely to meet the sign of the flying goose—the emblem of the National Wildlife Refuges. You may meet it by the side of a road crossing miles of flat prairie in the Middle West, or in the hot deserts of the Southwest. You may meet it by some mountain lake, or as you push your boat through the winding salty creeks of a coastal marsh.
>
> Wherever you meet this sign, respect it. It means that the land behind the sign has been dedicated by the American people to preserving, for themselves and their children, as much of our native wildlife as can be retained along with our modern civilization.

In 1999, the USFWS adopted a standardized image of the blue goose to appear on all national wildlife refuge signs. Since then, Puddles has appeared increasingly in a fuzzy, human-sized form. Playing a starring role at many National Wildlife Refuge System venues and events, Puddles receives a warm welcome from children and adults alike.

DEREGULATION, EXPANSION, AND COLLABORATION

The George W. Bush administration oversaw some of the most concerted attempts to roll back regulatory protections for public lands in general, and national wildlife refuges in particular. As part of the administration's national energy policy, efforts were made to facilitate oil, gas, and other resource development on the public domain. The president threw his full support to initiatives designed to open ANWR to offshore oil drilling, but they repeatedly came up short in Congress. The administration also engaged in numerous attempts to weaken ESA implementation by intervening in USFWS field reports and scientific recommendations, again with mixed results.

Yet in contrast to his stance on environmental regulation, President Bush, on the eve of leaving office, took an unexpected turn on refuge expansion. While only seventeen new refuges were established during his two terms (twelve by executive order), on January 6, 2009, President Bush took the audacious step of creating the single largest unit in the history of the national wildlife refuge system. On that day Bush evoked the Antiquities Act to establish the Marianas Trench Marine National Monument. Covering over 95,200 square miles in the South Pacific, the monument included three managerial units: the Trench, Volcanic, and Island Units. Ten days later, two of these units were transferred to the national wildlife refuge system as the Mariana Trench and Mariana Arc of Fire National Wildlife Refuges, respectively (figure 6.5). Both refuges consist primarily of submerged lands on the ocean floor. Nonetheless, environmental groups lauded the move. The Mariana Trench Refuge is over fifty million acres in size, encompassing the deepest underwater canyon on the planet. Meanwhile, the Arc of Fire National Wildlife Refuge consists of a cluster of 2.3-mile-diameter areas of protection centered on each of twenty-one undersea mud volcanoes and thermal vents.

To make sense of these seemingly contradictory actions, it helps to take a step back. It is worth noting, for example, that oil and gas exploration and development were not restricted under the executive orders creating the Mariana Trench Refuge. Hence, in this instance at least, there is no conflict of interest between the fossil fuel industry (one of President Bush's strongest political supporters) and refuge creation. Second, the new refuges are not only incredibly distant from most Americans—both geographically and mentally—but also, well . . . underwater. These attributes greatly reduce the likelihood of conflict with property rights groups and development interests. In short, their location and the inclusion of regulatory caveats both ensure that the refuges align with a nature-as-commodity philosophy. Currently, they do not appear to hold economic value, but if it turns out otherwise, the law allows for resource development.

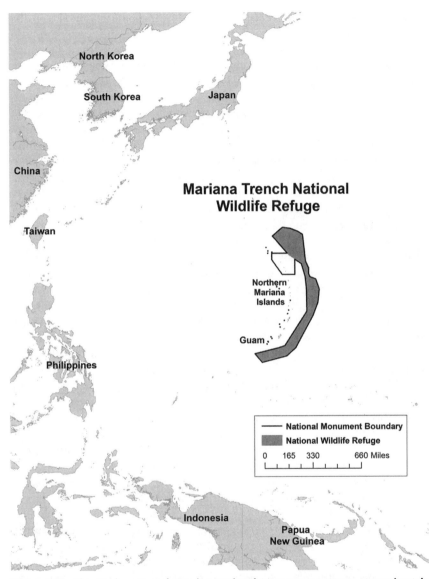

Figure 6.5. The Mariana Trench Marine National Monument protects approximately 95,200 square miles of submerged lands and waters in the Central Pacific Ocean and includes the deepest place on the Earth's surface. Map by Jessica Lee.

Finally, President Bush's action reflects the powerful appeal that the creation of monuments, refuges, or other forms of public land can hold for politicians. It is a relatively easy way for presidents to leave a positive legacy for themselves without having to deal with Congress or the thorny budgetary questions of how to pay for proper management. A late-term executive order offers immediate public goodwill while postponing more difficult budgetary debates until after the president leaves office. For this reason, history provides many illustrations of politicians—even those adamantly against federal regulations during their careers—turning "green" as they walk out the door. Such actions are perhaps politically expedient, but they also reflect to some degree the age-old notion of nature as unpeopled, eternal, and pristine. The focus is on drawing the boundary and less about what happens to the dynamic and interconnected webs of flora and fauna after the dedication ceremony.

Since 2009, the national wildlife refuge system has continued to grow and develop, but at a slower pace and in different ways. As mentioned, at the close of President Obama's first term, only five new refuges or conservation areas had been added to the system. This reflects, in part, the fact that many of the remaining large-scale tracts of land within the public domain suitable for designation as wildlife refuges have already been set aside. Unless one looks to unconventional locations (e.g., lands held by the Department of Defense, or upon the ocean floor . . .), the largest remaining tracts of wildlife habitat in critical need of protection are most likely to be found on privately owned lands. And when these areas consist of heavily fragmented parcels, conservation efforts can become even more complicated. Executive orders are of little utility here. Rather, new strategies are needed that allow USFWS officials to work together with landowners and land trust organizations, such as the Nature Conservancy, to protect wildlife habitat.

The National Wildlife Refuge Volunteer Improvement Act of 2010 represents one attempt to help establish a new public-private partnership paradigm. The law continues funding and expands existing volunteer and community partnership programs under the 1998 Volunteer and Community Partnership Enhancement Act. Specifically, it calls for a national strategy for the coordination and utilization of NWR volunteers and establishes a network of regional volunteer coordinators. According to the Act, over thirty-nine thousand volunteers regularly contribute more than 1.4 million hours of work annually to the wildlife refuge system.[34]

An excellent example of this collaborative approach is the Everglades Headwaters National Wildlife Refuge and Conservation Area established in Florida in January 2012. It was designed to protect water resources in the Everglades Headwaters, sensitive grassland and savanna habitats, and maintain rural working landscapes based on ranching and farming.[35] It is comprised of a 150,000-acre area, consisting primarily of privately owned lands. Within these boundaries, the USFWS is authorized to buy land or conservation easements from willing landowners within an identified fifty-thousand-acre zone, and to purchase conservation easements only in a larger one-hundred-thousand-acre zone. Only lands purchased outright may become part of the wildlife refuge. The program is therefore voluntary and is funded through the Duck Stamp program and the federal Land and Water Conservation Fund. Interestingly, the refuge formally came into being after the Nature Conservancy donated ten acres to the USFWS. Upon completion, the Headwaters Refuge and Conservation Area will comprise a blended tapestry of private, public, and conservation easement lands that protect wildlife while also allowing the continuation of human residents and "working landscapes."

CASES

Despite the new advances in collaborative approaches to wildlife management, the continued (and sometimes dramatic) expansion of the system, the strengthening of the agency's conservation mandate, and creation of new volunteer programs, the national wildlife refuge system still faces many challenges. For example, today the system remains weighted—in terms of acreage—in places well beyond the reach of most Americans. As a result, the public is still largely in the process of becoming acquainted with the nation's refuges. In the spirit of Smokey Bear, the National Wildlife Refuge System's Puddles the Blue Goose campaign is a step in the right direction, but there is still a long way to go. Despite the media attention, the Mariana Trench designations were akin to the creation of the Alaskan refuges. President Bush's order dramatically expanded the refuge system, but it did so in a place that most Americans may not easily identify on a map, much less visit.

However, one of the most effective ways to cultivate ties with the broader public appears to be the investment in federally supported volunteer and partnership programs. The continued cultivation of long-held relationships with organizations such as the National Wildlife Federation, the Audubon Society, and Ducks Unlimited aligns perfectly with such initiatives. In addition, the move toward collaborative approaches in both management and land acquisition will provide increasing opportunities for the practical expression of these relationships in the service of wildlife protection.

Nonetheless, other challenges remain, from balancing recreation, hunting, and conservation management priorities to fending off resource development initiatives inside the refuges themselves. Endangered species protections and coordination with state wildlife agencies also continue to pose difficulties for the USFWS. Many of these issues trace their roots to the continued influence of problematic conceptions of nature, as illustrated in the two case studies below. The case of ANWR underscores the persistent influence of nature-as-commodity thinking, which has fueled the decades-old debate over oil and gas drilling in the region. Meanwhile, the story of the National Elk Refuge is a classic example of how difficult it is to shake off the notion of the public domain as "unpeopled, eternal and pristine," with potentially devastating effects for both humans and wildlife.

Arctic National Wildlife Refuge

Perhaps the most famous unit in the national wildlife refuge system is the Arctic National Wildlife Refuge (ANWR) in Alaska (figure 6.6). Spreading over 19.2 million acres in the northeastern corner of the state, it serves as the largest land-based refuge in the nation. The landscape is extremely diverse, ranging from the boreal forest and nine-thousand-foot peaks of the Brooks Range in the south to the broad coastal plain and Arctic Sea in the north. The massive refuge contains habitat for a myriad of marine and terrestrial species, including polar bears, golden eagles, and the famous Porcupine caribou herd. With seasonal migrations across the U.S.-Canadian border, the Porcupine herd comprises the second largest elk herd in North America and the seventh largest in the world.[36] Each year, powerful winds off the coastal plain provide relatively insect-free calving grounds for the herd.

First designated in 1960 as the 8.9-million-acre Arctic National Wildlife Range, this vast area in the northeastern corner of Alaska was expanded by 10.3 million acres and reclassified as the Arctic National Wildlife Refuge with the passage of ANILCA in 1980. In so doing, the Act simultaneously designated some eight million acres of the refuge as wilderness, and placed 1.5 million acres of the coastal plain into a unique category. In particular, Section 1002 of ANILCA called upon the Interior Department to assess the potential for oil and gas development in the designated zone, but it also prohibited such development unless authorized through an act of Congress. And herein lies the rub.

The Interior Department completed its first report in 1987, and the battle has raged ever since. On one side is the Alaska Coalition, a massive group of environmental NGOs, local activists, wildlife ecologists, and the Gwich'in Eskimos, indigenous peoples who have traditionally relied on the Porcupine caribou herd for their subsistence. On the other side is the Coalition for American Energy Security, consisting of representatives from the oil and gas industry, as well as motorized recreation,

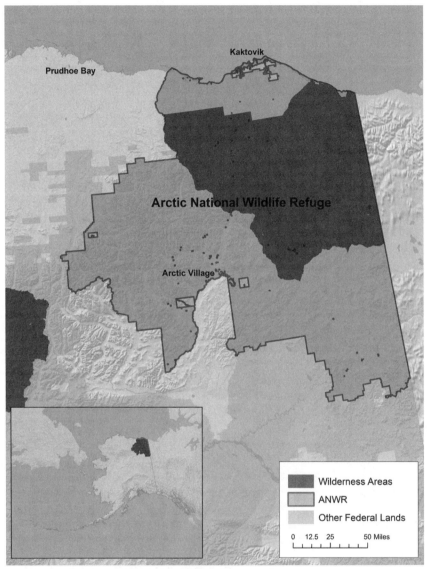

Figure 6.6.   The Arctic National Wildlife Refuge. Map by Jessica Lee.

agriculture, land developers, and Inupiat Eskimos, the only indigenous peoples actually residing within the boundaries of ANWR. Centered in the town of Kaktovik, the Inupiat also control over ninety-two thousand acres of subsurface mineral interests in the region.

Meanwhile, the Alaskan state government and the majority of state residents heavily support oil development, a stance that strongly reflects the state's profound financial dependence upon the industry. As part of a unique statehood agreement, Alaska was allowed to retain 90 percent of royalty revenues from oil development within the state. Since statehood, these funds have provided between 85 percent and 90 percent of state revenues. As a consequence, Alaska collects no state income tax, but it instead pays out annual dividends to each citizen. Since the initial payment of $1,000 to every eligible man, woman, and child in 1982, the amount has fluctuated from a low of $331 in 1984 to a high of $2,069 in 2008.[37] The local political influence of Alaskan oil simply cannot be overstated.

At first glance, the arguments on each side appear to reflect the classic debate between two competing visions of nature: nature-as-commodity and nature as an entity with inherent value. Closer inspection, however, reveals that both sides tend to emphasize the relative value of ANWR in commodity terms. The focal point for much of the debate consists of differing interpretations of the data. Since government studies provide only a range of estimated costs and benefits, including how much oil exists (between 600 million and 9.2 billion barrels, according to the 1987 study) as well as the resultant economic and environmental impacts, each side frequently points to opposite sides of the spectrum to bolster their positions.

Advocates for oil development consistently offer a variety of well-rehearsed rationales. First, they argue that there is a vast amount of oil in the region and failure to develop it would be a colossal waste of natural resources. Second, they assert that the oil is needed for national security. It will not only lessen U.S. dependence on Middle Eastern oil (and hence, our entanglements in Middle East conflicts), but will also help reduce the federal deficit. Consumers will enjoy lower gas prices, while many more will benefit from increased employment opportunities and overall economic growth. More recently, arguments emphasize that with recent technological advances, the footprint of ecological impacts resulting from resource development will be extremely minimal. They claim that less than 1 percent of the entire coastal plain will be affected, rendering no significant harm to the flora and fauna within the refuge.

Conversely, environmental groups argue that the quantity of oil is insubstantial (less than two hundred days' worth, according to a 1991 study) and certainly *not* worth the negative environmental impacts that are conceived as much larger than industry estimates, despite the advent of new technologies. They are also concerned that the oil extracted from ANWR may be exported as a function of global supply and demand, rather than consumed domestically.[38] Moreover, regardless of U.S. production, the Middle East still plays the most influential role in setting global prices and serves as a major supplier to U.S. allies in Europe and around the world. Finally, despite the relatively small total footprint of development, they argue that its spatial configuration will stretch over hundreds of miles, thus creating a disproportionate impact on a uniquely fragile ecosystem. Hence, development of ANWR oil is not likely to produce the economic or political benefits claimed by the oil industry. Rather it presents an effort to "destroy America's Serengeti" for an inconsequential amount of oil.[39]

The political battle over ANWR kicked off with lawsuits filed by environmental groups over the Interior Department's 1987 findings and recommendation to Congress to open the refuge to oil and gas leasing. Meanwhile in Congress, prodevelopment Republicans pushed legislation to open the refuge to oil drilling, while conservation-minded politicians (often, but not always, Democrats) offered bills that would give ANWR wilderness protection. Time after time, bills were derailed or amended to remove specific language pertaining to ANWR. With the arrival of George H. W. Bush in 1989, support in both the House and Senate suggested a law would finally be passed to allow drilling. However, the Exxon *Valdez* oil spill in Prince William Sound underscored environmentalist arguments that oil development carried profound risks. Once again, the matter was set aside.

The advent of the Gulf War in 1990 reopened the discussion of the national security need for increased domestic oil production. Support was bolstered by a 1989 Bureau of Land Management study that raised the estimated amount of oil on the Coastal Plain to between 697 million and 11.7 billion barrels. However, in 1991, once again public opinion turned against the oil industry. A *New York Times* story reported historically high quarterly profits for U.S. oil companies—ranging from 35 percent to over 68 percent—as they engaged in profiteering in the ramp-up to the Gulf War.[40] Meanwhile, increased production by Saudi Arabia resulted in an oil glut on the world market. Once again, legislative efforts to open the refuge were stymied.

During the Clinton administration, a 1995 Interior Department study again revised oil estimates in ANWR, this time lowering the numbers to between 148 million and 5.15 billion potential barrels. Nonetheless, with Republican majorities in both houses of Congress and Alaskan politicians in charge of the two most important congressional committees, an amendment to open ANWR was included in the 1996 omnibus budget bill. Significantly, budget bills cannot be filibustered. With passage in both the House and the Senate, it seemed ANWR's fate was finally sealed until President Clinton vetoed the bill. His action effectively shut down the federal government for a week. Though not the only reason for his veto, Clinton cited ANWR as one of the major rationales for taking action. When Congress finally did pass a budget bill for 1996, it no longer included mention of ANWR.

Over the next decade, the battle continued. During the George W. Bush administration, bills continued to be offered, but most failed in the Senate. The closest call came in 2005. That year, each of three bills—the Energy Bill (April), a federal budget resolution bill (March), and the defense appropriations bill (December)—included provisions allowing ANWR to be opened to drilling.[41] In the first two cases, bills passed both the House and Senate, but ANWR amendments were stripped during conference committee negotiations. The third attempt fell victim to a successful Democratic filibuster. In 2008, President Bush tried one last time to push Congress to open ANWR and lift the general ban on offshore oil drilling, but once again, his effort fell short.

After taking office in 2009, President Obama declared his intention to protect ANWR from oil development, even as he agreed to allow the resumption of offshore oil drilling elsewhere. Some environmental advocates now look to congressional wilderness area designation as the best chance for permanent protection for the refuge. But the partisan politics surrounding the issue continue to present an imposing barrier. Indeed, the ANWR debate is expected to endure as long as the notion of nature-as-commodity continues to provoke such strong sentiments in discussions over wildlife management priorities in the United States.

The National Elk Refuge

Another manifestation of this conceptual legacy is found in the debates over the National Elk Refuge, located outside the town of Jackson Hole, Wyoming (figure 6.7). Established in 1912, the National Elk Refuge was born of the same spirit that created refuges for bison in Montana and Oklahoma. By the early twentieth century, elk herds once numbering in the tens of millions in North America had dwindled sharply to approximately fifty thousand.[42] In Wyoming, one of the largest herds, totaling some twenty thousand animals, fell under intense pressure from hunters, trappers, and the loss of habitat from homesteading ranchers and their livestock. In particular, growth of farms and ranches outside the town of Jackson came to occupy some 75 percent of the elk herd's winter range. While the herd typically spent late spring, summer, and early fall in the Rocky Mountain high country, winter snows typically forced them to lower-elevation valleys to graze. But human settlement in those same valleys put the continued viability of the herd in question.

Figure 6.7. Winter on the National Elk Refuge near Jackson Hole, Wyoming. Photo courtesy of the USFWS.

The National Elk Refuge was designed to restore the herd by ensuring continuous winter range. In the first year, a particularly fierce winter led managers and local residents to purchase hay to supplement natural forage and stave off the threat of starvation.[43] The strategy was deemed a great success as the number of elk began to rebound. However, rather than return to natural forage the following year, the practice of supplemental winter feeding was soon adopted as standard practice. Meanwhile, as the elk herd grew, so, too, did the refuge. Through donations from the Izaak Walton League, presidential executive orders, and congressional appropriations for the purchase of private lands, the refuge expanded from its original several thousand acres to over 24,700 acres. Currently, the refuge provides annual winter range for three thousand to ten thousand elk, and year-round habitat for a myriad of other species. Coordinated management with nearby Grand Teton National Park and contiguous national forestlands provide additional protection for a portion of the herd's larger summer and winter range. Meanwhile, the state of Wyoming has established twenty-two other winter feeding grounds for elk scattered around the state.

Debate centers on the question of artificial human feeding of the elk, both on the National Wildlife Refuge and the other state-run elk feeding grounds in Wyoming. Environmental advocates contend that this practice has led to an overpopulation of the elk herd, one that no longer sustainably resides on the refuge given its size and available resources. One potential outcome of this overcrowding is increased risk for devastating diseases, including hoof rot, brucellosis, and most importantly, chronic wasting disease, a condition that causes neurological damage and death in infected animals. Chronic wasting disease cases have been found in wildlife populations across the United States, with most occurring in feed-lot conditions. Once established, the disease-causing agents (called prions) can persist in soil for decades. To date, none of these diseases have been detected on the refuge. Nonetheless, wildlife advocates and ecologists attest that the overcrowded conditions represent a profound health risk in terms of disease transmission if an outbreak should occur. In a worst-case scenario, they argue, the refuge would have to be shut down.

On the other side of this debate, local ranchers worry that unfed elk will venture onto their cattle pastures and compete for forage on publicly leased grazing lands. The local tourism industry points to the thousands of people who pay good money to ride in sleighs through vast elk herds on the refuge, and of course, stay in local hotels and eat in restaurants.[44] Meanwhile, hunting guides and outfitters are concerned that they will lose business from people who fear that fewer elk on the refuge will translate into failed hunting expeditions.

The USFWS addressed this issue in the 2007 Elk Refuge Management Plan, in which they decided to allow for the indefinite continuation of supplemental feeding. However, in 2008, the environmental NGO Earthjustice filed a lawsuit on behalf of the Greater Yellowstone Coalition, Jackson Hole Conservation Alliance, Defenders of Wildlife, Wyoming Outdoor Council, and the National Wildlife Refuge Association. The suit accused the USFWS of failing to fulfill its mandate to protect wildlife as stated under the 1997 National Wildlife Act.[45] In August 2011, a federal appeals court ruled that the USFWS must end the practice of supplemental feeding. Specifically, the court stated that "[t]here is no doubt that unmitigated continuation of supplemental feeding would undermine the conservation purpose of the National Wildlife Refuge System."[46]

One of the great ironies of this conflict is the way it contrasts with the issue of bison management in the Yellowstone region. Some residents who volunteer to help feed elk on the refuge or other winter feeding grounds simultaneously maintain strong opposition against the presence of bison outside the boundaries of Yellowstone National Park (see chapter 4). Concern that bison might carry and transfer brucellosis to domestic livestock marks them as an eminent threat to the ranching industry.

However, such sentiments have not yet carried over to elk, despite the fact that overcrowding on the National Elk Refuge may lead to increased incidences of both brucellosis and the even more deadly chronic wasting disease in the region.[47]

The case illustrates the continuity of several broad underlying conceptualizations of nature. One of these stems from the national wildlife refuge system's long history of association with the sport hunting community. From the 1894 Yellowstone Wildlife Protection Act to the various iterations of the Duck Stamp Act, hunting organizations have played a powerful role in both supporting and shaping wildlife laws. Their influence continues to be felt as the USFWS grapples with the challenge of balancing recreation and conservation management priorities. At a fundamental level, the recreation mandate is closely tied to a nature-as-commodity perspective. In debates over the National Elk Refuge, it is hard to ignore the economic arguments put forth to prioritize game species at the expense of others. It is not just that hunting is fun, but rather, that it renders economic benefits to local communities.

Finally, the case demonstrates the persistent effects of viewing nature as eternal and pristine. Approximately one hundred years ago, in order to protect the elk, the decision was made to build a refuge. According to this logic, demarcating a boundary meant the elk were now a problem largely solved. Once safely encompassed and protected from outside threats, "pristine" nature would stabilize and thrive in perpetuity. But as was the case in the early days of Yellowstone National Park, this logic failed to take into account the dynamic and systemic aspects of ecology. Nature—even nature protected by a refuge—is constantly changing in response to changing conditions. And these conditions can be altered by actions and events taking place both *within* and *beyond* the borders of a refuge. In this case, efforts by local residents and USFWS managers to extirpate predators and provide an "unnaturally abundant" amount of winter forage have allowed elk populations to expand well beyond the land's carrying capacity. At the same time, the growing density of the elk herd may be creating conditions for increased risk for disease outbreaks that threaten both the wildlife and the domesticated livestock in the region.

Dealing with the legacy of these conceptual assumptions and past management decisions adds to the complexity of the task facing refuge managers today. As the USFWS currently works to craft a new comprehensive management plan for the National Elk Refuge, the challenge of breaking free of these problematic conceptions of nature remains. Their influence continues to be felt as managers and stakeholders search for that elusive balance between various recreation and resource development activities and the implementation of a meaningful wildlife conservation mandate.

7

Bureau of Land
Management Lands

The plan is to get rid of public lands altogether, turning them over to the states, which can be coerced as the federal government cannot be, and eventually to private ownership. . . . The immediate objectives make this attempt one of the biggest land grabs in American history. The ultimate objectives make it incomparably the biggest.

—Bernard DeVoto (1947), on the attempt by western stockmen
and politicians to privatize the public domain

In all my 20 years in the U.S. Senate, I have never seen a clearer example of the arrogance of federal power. Indeed, this is the mother of all land grabs.

—Senator Orrin Hatch (1996), on the declaration of Grand
Staircase-Escalante National Monument by President Clinton

Despite the passionate statements noted above, if there is one type of federal land that is less recognized and understood by the public than wildlife refuges, it has to be the lands administered by the Bureau of Land Management, or BLM. Historians frequently adorn BLM lands with colorful names, such as "the lands that nobody wanted" or "the leftover lands." To this list might also be added "the lands that no one ever heard of."[1] Such a claim, of course, depends greatly on where you live and how you make a living. For example, residents of the rural West, especially those in the ranching and mining business, are often quite well acquainted with BLM lands as sources of pasture or desirable mineral deposits. Western state governments know them as lands beyond their

179

jurisdictional reach, but which may constitute significant portions of state territory. Nonetheless, for many Americans, both the BLM and the public lands it manages remain nonentities.

What makes this striking is that BLM lands constitute the largest portion of the public domain. With over 247 million acres (and another 700 million acres of subsurface mineral deposits),[2] BLM lands are more than three times the size of the national park system, nearly a third larger than the national forest system, and encompass over one hundred million acres more than the national wildlife refuge system. Moreover, they are the *only* lands legally termed *public* in the federal land system.

So what are BLM lands? Where are they located? What types of landscapes do they contain? What is their primary purpose? And what makes them so contentious? Answering these questions takes us to the heart of the larger issue regarding the disconnect between BLM lands and the U.S. public. It also underscores the powerful conceptual legacy of seeing nature as both "empty" (if not necessarily pristine or eternal) and as a commodity.

Of all the different types of federal lands, BLM lands were among the last to be so designated by the federal government. They represent the portion of the public domain *not* selected for homesteading, *not* sold off, and *not* given away in the form of land grants to railroads or other private interests during the period of disposal (chapter 2). Neither were these lands chosen for classification as national parks, forest reserves, wildlife refuges, or other types of protected areas, including federal Indian reservations or military reserves (chapter 3). In short, they constitute all remaining portions of the public domain after the lion's share of public land sales, grants, and designations came to a close in the early twentieth century.

The lack of public awareness undoubtedly stems from this fact. But there are other reasons. As with the national wildlife refuge system, geography explains much. Like the refuges, BLM lands lie almost exclusively in the eleven most western contiguous states and Alaska (figure 7.1). In fact, the BLM controls about 23 percent of the total surface land area of the region.[3] In some states, such as Nevada, BLM lands account for nearly 68 percent of state territory. This strong western focus quickly diminishes the relevance of BLM lands for those residing east of the Rocky Mountain states.

Add to this the fact that many BLM lands consist of a spotty patchwork of parcels, and the picture gets even more complicated. The alternating checkerboard pattern of public and private lands in much of the Southwest and Alaska creates confusion for the general public and tremendous challenges for federal land managers, whose priorities may be at cross purposes with those of adjacent landowners.

Bureau of Land Management Lands

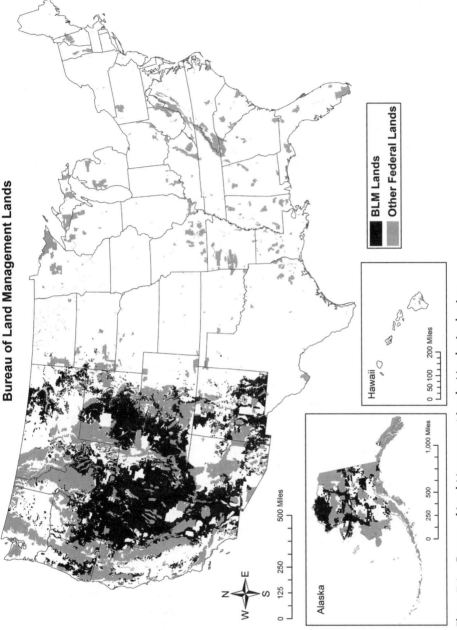

Figure 7.1.   Bureau of Land Management Lands. Map by Jessica Lee.

Second, BLM lands comprise some of the most arid landscapes on the continent. Claiming much of the Great Basin (the vast expanse between the western slope of the Rockies and the Sierra Nevadas) and large swaths of the desert Southwest, BLM lands encompass considerable portions of the seemingly barren low-lying territory from the Colorado plateau to the Columbia River basin. In terms of resource development, such lands are often best suited to livestock grazing and mining rather than agricultural production, logging, or serving as vacation destinations. Thus, while many people may pass *through* these landscapes en route to Yellowstone, the Grand Canyon, Las Vegas, or elsewhere, few stop to consider them as part of the same public land heritage as national parks.

A third reason for the lack of public understanding has to do with management priorities. On a continuum ranging from environmental protection to resource development, BLM lands weigh heavily on the development end of the scale. Historically, the dominant industries have been livestock grazing and mining, though logging has also played a significant role on some BLM lands, especially in the Pacific Northwest.[4] Absent a tradition in recreational use or environmental preservation, the general public has largely avoided BLM lands as a vacation destination—or for that matter, a destination of any kind. For those who do pass through, the sight of barbed wire fencing suggests the land to be privately owned rather than public. Certainly, this view is promoted by more than a few grazing permit holders. Given this commodity focus, some environmentalists argue that "BLM" actually stands for "Bureau of Livestock and Mining."

But the BLM is not alone in its resource development emphasis. The bureau actually adheres to a multiple-use approach that most closely resembles the USDA Forest Service than any other public land agency. Like the national forests, BLM lands were born out of crisis. In place of the threats posed by forest fires and floods, BLM lands were established largely in response to overgrazing and the combined stresses of the Great Depression and the Dust Bowl in the 1930s. Both agencies were founded on Progressive Era ideals of multiple use and science-based management designed to protect natural resources from the ravages of unfettered commercial development. According to this perspective, public lands are not to be preserved, but used. Resource development is the first priority, but done in a way that reduces waste, increases efficiency, and ensures the continuation of resource supplies for future generations. For the Forest Service, putting these ideals into practice meant maintaining close ties to the logging industry. For similar reasons, the BLM has long been associated with livestock production and mining.

However, unlike the Forest Service, which increasingly developed a much broader recreational and conservation-minded constituency after World War II, the BLM held to its industrial focus until much more recently. And the differences don't end there. Recall that the Forest Service was the beneficiary of a distinct regulatory mandate, a potent esprit de corps, and a tradition of professional, scientifically based management protocols, all shaped and put into place by a charismatic and influential leader. The BLM lacked all of these things. No Gifford Pinchot or Stephen Mather helped chart the way forward for the young agency. Rather, the BLM suffered from the beginning from weak regulatory powers, mixed congressional support, and an administrative structure overly influenced by the very industry it was intended to monitor.[5]

In more recent years, recreation, ecological conservation, and wilderness preservation have grown in importance as management priorities. Like other public land agencies, this change has been slow in coming, though arguably the BLM set new standards for sluggishness. Nonetheless, today the BLM manages a number of national monuments, recreation areas, and wilderness areas within its jurisdiction as part of the agency's National Conservation Lands System. These account for over 10 percent of the lands and resources under the BLM's care.

Finally, there is no Old Faithful, Smokey Bear, or even Puddles the Blue Goose here. Indeed, the best chance for public awareness and goodwill— BLM mascot Johnny Horizon—was unceremoniously decommissioned in 1976 (see below). Aside from Grand Staircase-Escalante National Monument, it is difficult to find an iconic landscape, figure, or symbol to draw public attention, bolster visitation, or underscore core management priorities for the BLM, nothing that powerfully reflects or refutes the conceptual values and assumptions underpinning the "leftover lands." And this in itself may be the most telling characteristic of BLM lands. They are the "void": the portion of the public domain whose apparent emptiness stands as its most powerful moniker.

This perceived lack of value, the absence of identifiable economic, aesthetic, or recreational worth, best explains not only the lack of public awareness today but also the reason they were passed over by homesteaders in the first place, rendering them available for classification as BLM lands. However, like the story of Yellowstone, the converse is also true. Their acknowledged commodity value as forage and mineral deposits is precisely what has driven certain industrial interests to fight long and hard against federal regulatory control. As a consequence, the conceptual legacy of seeing nature as a commodity is perhaps most evident in BLM lands than in any other federal land type. This seeming contradiction does much to explain the historical evolution of these lands and the BLM itself.

RETHINKING THE UNWANTED LANDS:
JOHN WESLEY POWELL

The public lands managed by the BLM represent the final act in the story of land disposal and privatization that dominated the nineteenth century (chapter 2). Recall that since 1812, the General Lands Office oversaw the transfer of the public domain into private hands, whether as land grants to states, canal or railroad companies, preemption entitlements, mining claims under the 1872 Mining Act, or individual settlement claims made under the various homestead acts.

Despite the exhaustive national effort to sell off the public domain, there were numerous calls by early environmental conservationists to reconsider this path. While national park advocates such as John Muir focused their efforts on the most spectacular and "sublime" landscapes (Yosemite Valley, Mount Rainier, Grand Canyon), and progressive conservationists such as Gifford Pinchot and Teddy Roosevelt looked to set aside forested landscapes or wildlife sanctuaries, few gave much heed to the arid and presumably "useless" desert lands of the interior West.

The one nineteenth-century exception to this rule was John Wesley Powell. A one-armed veteran of the Civil War, Major Powell famously completed the first recorded voyage down the Colorado River through the Grand Canyon by a Euro-American. His expedition was sponsored by the 1876 appropriations bill for the General Lands Office, which mandated surveys of the public domain to discern the best use and classification of lands under the various homestead acts. Knowing which lands were rich with mineral, timber, or water resources (and which were not) was crucial for federal officials. It allowed them to guard against speculators seeking to buy resource-rich lands for prices well below their value by claiming them to be worthless desert lands.[6] To this end, the government dispatched expeditions by Ferdinand Hayden, George Wheeler, Clarence King, and John Wesley Powell to explore various parts of the West.

However, unlike Hayden's famous visit to Yellowstone, Powell's explorations did not result in a new national park or the retention of specific lands by the national government, but rather a complete rethinking of how the arid West should be settled altogether. Upon his return to Washington, Powell issued two reports. The first, published in 1875, was titled *Exploration of the Colorado River of the West and Its Tributaries*. The document received widespread public attention for its action-filled depictions of Powell's men shooting the rapids and other adventures encountered while traversing the Grand Canyon. However, it was Powell's second work, *Report on the Lands of the Arid Region of the United States*, published in 1879, that laid out a fundamentally new—and controversial—way of understanding the desert landscapes of the Southwest.

Highlighting the dry conditions and scarcity of water resources found there, Powell argued that the region could support only a limited number of settlers. Such claims directly challenged the "rain follows the plow" theories espoused by local boosters at the time. Hence, it was no surprise that powerful land speculators seeking to attract settlers into the region quickly marked Powell as a political adversary.

Powell's arguments were threefold. First, he pointed out that settlement required irrigation. The standard 160-acre homestead farm, however, was simply too large an area for a single family to manage. He therefore recommended a reduction to eighty acres for such farms. On the other hand, for livestock producers, a much larger area was needed. Not only was the standard 160-acre homestead too small, but so, too, was the 640 acres made available under the Desert Lands Act of 1877. Given the meager forage available in the region, Powell suggested viable ranching operations would need parcels of at least 2,560 acres. And all lands used for grazing should not be fenced but held in common by neighboring ranchers to allow for the most efficient use of available pasture.

Second, he argued that the rectilinear survey system used elsewhere in the United States to carve out homestead units was not appropriate for demarcating claims in arid regions. Rather, he promoted the idea of using natural ecological boundaries—ideally, watersheds—to delineate property boundaries. This idea, which resonates with the concept of ecosystem management, would ensure access to scarce water resources for all settlers.

Finally, recognizing that the cost of irrigation works was beyond the financial reach of most individual farming families, Powell advocated federal government involvement in the development of cooperative water districts. In this way, the necessary water retention and distribution systems could be constructed, while remaining accessible to all homesteaders.

Ultimately, most of Powell's recommendations went unheeded by Congress. Nonetheless, with backing from the National Academy of Sciences and Interior Secretary Carl Schurz, Powell successfully crafted congressional legislation resulting in the creation of the U.S. Geological Survey (to oversee the comprehensive mapping of the nation's lands and resources) and the Public Lands Commission (to provide direction to Congress on western settlement). The Commission's 1880 report adopted several of Powell's ideas, such as scrapping the rectilinear survey system, but it had little influence otherwise.[7]

It would take almost two decades before some of Powell's recommendations gained federal acceptance. But it finally happened—at least to some degree—with the passage of the Carey Act in 1894 and then again, with the creation of the Bureau of Reclamation in 1902. The Carey Act transferred federal lands to states and private water companies to help finance irrigation works to encourage settlement in arid regions. Meanwhile, the Bureau of Reclamation carried out large-scale federal dam and irrigation projects throughout the West. The 1902 law included a "Powell-esque" requirement that water produced from these works be used exclusively by small-scale family farms from 160 to 320 acres in size. However, this provision was essentially ignored until 1982, at which time Congress removed it from the law altogether. Thus, in the absence of significant public land law reform, land use on western deserts and plains continued down a path that could only lead to crisis.

TRAGEDIES OF THE (RANGELAND) COMMONS

For much of the nineteenth century, mining and ranching comprised the dominant land uses in the more arid regions of the public domain. From 1849 to the 1860s, the gold and silver mining rushes in California, Colorado, Nevada, and elsewhere took place mostly in the high country of the Sierra Nevadas and Rocky Mountains. However, mining camps served as catalysts for other economic activities, driving demand for transportation infrastructure, processing facilities, and all manner of merchandise for miners and settlers alike. As one of the most sought-after goods, beef provided an impetus for livestock production on nearby rangelands to serve local mining markets.

After the Civil War, with the arrival of transcontinental railroads, the western cattle industry quickly evolved into a massive agribusiness. The largest operations began in Texas. Raising calves on the public domain with forage free for the taking, livestock producers drove herds north to railroad lines in Kansas, Nebraska, or other western depots for transport to eastern markets. With such low production and transport costs, profit margins were enormous and soon attracted investors from the East Coast, and increasingly, from Great Britain.[8] The influx of foreign capital radically expanded the size and scale of operations across the region.

In terms of acreage, the livestock industry wielded a tremendous impact on the public domain. Making liberal use of the homestead acts, ranchers made claims strategically designed to control limited surface and groundwater sources. Without access to water for irrigation, the remaining public rangeland became useless for new settlement, discouraging homesteaders, migrant sheepherders, and burgeoning rival ranching outfits. Instead, the land was transformed into de facto private pasture for established livestock operations. In this way, ranchers could control tens of thousands of acres of land, while holding title to only several hundred. Soon the size of the largest operations surged well beyond the 2,560-acre limit suggested by Powell.[9]

By the 1880s, the western ranges were overstocked. Domesticated cattle essentially replaced the native bison herds on the plains. In the fight against incursions onto "their" rangelands, ranchers made increasing use of a new invention: barbed wire. This inexpensive fencing material allowed ranchers to illegally close off portions of the public domain to settlers, sheepherders, and rival stockmen. Meanwhile, stretched beyond their carrying capacity, public rangelands suffered the effects of overgrazing and soil erosion.

In 1886, a dry spring and summer reduced forage and water availability still further, decimating the health of hundreds of thousands of cattle on the plains. Then winter arrived. Amid freezing temperatures, weakened herds died by the millions, often stacked up on one another against barbed-wire drift fences. As the snows melted away, ranchers tallied the cost. Some lost everything, but the effects were uneven. Because of poor record keeping, historical estimates of the total impact vary wildly, ranging from 30 to 90 percent of all domestic cattle on the public domain.[10] Known as the "Big Die-Up," the image of denuded and eroded rangelands covered with the dead carcasses of domestic livestock painted an all-too-familiar picture in American history: unregulated resource use leading to a literal tragedy of the commons.

In the aftermath of crisis, the era of large-scale, open-range cattle drives came to a close. But even before the losses sustained by the Big Die-Up, the effects of overstocking had already rendered difficult economic consequences. The increased supply drove prices down below the threshold of profitability for many producers, making them vulnerable regardless of environmental degradation on the plains.

At the federal level, Congress responded to the crisis and other problems of public land law abuse by passing the 1891 General Revision Act. Repealing or amending some of the most influential homestead laws, the act also gave the president power to set aside forest reserves (chapter 5). Significantly, several of these reserves were established with provisions granting specific protections to grazing lands. In addition, Congress acted to expand the national park system, establish some of the first wildlife protection laws, and pass legislation supporting water development in the West (chapters 3 and 4).[11]

However, by the dawn of the twentieth century, the tide had turned once again. A series of relatively wet springs and summers allowed ranching to rebound, albeit no longer under the open range model. Wet conditions also persuaded many new immigrants to attempt settlement on the western plains. The First World War opened new export markets for American agricultural products, providing additional incentives for settlement and cultivation. Although Congress took some progressive action during this period, passing the 1920 Minerals Leasing Act, for example,[12] it also issued a new round of homestead laws to stimulate settlement and resource development on the public domain. These included the Enlarged Homestead Act of 1909 and the Stock-Raisers Homestead Act of 1916. The latter was designed to encourage newly expanded livestock operations on public lands by offering larger 640-acre plots. The new parcels required ranchers to carry out rangeland improvements, but few followed through and the stage was set for a new round of crisis on the rangeland commons.

By the late 1920s, the public rangelands were again filled beyond capacity with domestic livestock. Meanwhile, homesteaders were moving with increasing numbers onto the plains. Relying on rainfall for irrigation and using deep cultivation methods, they systematically removed native vegetation and reduced soil moisture. Finally, drought conditions returned to the region, setting the stage for a perfect environmental storm.

The swing toward federal deregulation reached its zenith in 1929. In the aftermath of the Teapot Dome and Elk Hills scandals,[13] President Herbert Hoover proposed the wholesale transfer of all remaining unclassified public lands to the states, less the subsurface mineral rights. Interestingly, western political leaders rejected the offer on the basis that without the associated mineral rights, the deal was not favorable. Overgrazed, arid landscapes were seen more as a burden than a boon. Perhaps they saw the writing on the wall, for later that year the stock market crashed, followed by a series of massive dust storms lasting through 1936.

Collectively called the Dust Bowl, dry winds lifted tons of fine soil from the western Great Plains into the atmosphere, blocking out the sun before burying houses and farmsteads. The dust spread east to the nation's capital, even reaching across the Atlantic to the shores of Western Europe.[14] Coupled with the onset of the Great Depression, the combined economic and ecological crisis on public rangelands could not have been more dire. The time for decisive action had come.

THE TAYLOR GRAZING ACT OF 1934

For many of the largest livestock operators on the public domain, the idea of accepting some federal oversight in exchange for assistance finally seemed palatable. Nonetheless, concerned that this oversight might come from the Forest Service, ranchers and western politicians opted for new legislation that might be better shaped to their advantage.

At the time, the only meaningful regulation of grazing on public lands occurred on national forests. In 1906, Gifford Pinchot put into place a permit system that charged ranchers a small fee per animal unit month (AUM), a measure of how much forage an adult cow with calf would eat in an average month.[15] The Forest Service also held the right to determine the number of livestock and length of time allowed for grazing in a given area. Ranchers fought vehemently against this arrangement, indignant that they should be required to pay for the right to use national forest grasslands.[16] Although the fee imposed was well below market price for comparable pasture on private lands—five cents per AUM for cattle and one cent for sheep—they only conceded to the system after realizing that restricted access might protect their own interests.[17] Therefore, stockmen and their representatives in Congress intended that any *new* regulatory framework for the remaining public rangeland give them much greater control of the process.

Of course, these actions were only part of a much larger national response to the crisis. The Franklin D. Roosevelt administration put into place a long list of programs under the auspices of his New Deal, all designed to get the country back on its feet. To improve farming techniques and land use practices, FDR created the Soil Conservation Service[18] and established local conservation districts. Some programs combined conservation goals with provisions to address unemployment, such as the Civilian Conservation Corps. But for public rangelands, the most significant congressional action was the Taylor Grazing Act of 1934.

Tellingly, the act was sponsored by Senator Edward Taylor (D-Colorado), a staunch critic of the Forest Service and its grazing policies. Hence, while the law was intended to "preserve the land and its resources from destruction or unnecessary injury," it also sought to "provide for the orderly use, improvement, and development of the range" and to "promote the highest use of the public lands . . . [that are] chiefly valuable for grazing and raising forage crops."[19]

Reminiscent of the National Park Service Act, the dual mandate offered by the Taylor Act put support for the livestock industry front and center—a goal seemingly at odds with the notion of improved rangeland conservation. For FDR's Interior Secretary Harold Ickes, long a conservation advocate, these contradictions rankled. Nonetheless, both he and the administration supported the compromise bill as the best chance of bringing the remaining public rangelands under federal protection.

The tensions inherent in the dual mandate played out in a number of ways. The original 1934 Act established grazing districts in approximately eighty million acres of "vacant, unappropriated or unreserved" rangeland on the public domain, excluding Alaska. In so doing, it effectively closed off the land to settlement under the Homestead Acts.[20] To manage the districts, the Interior Department set up a permit and fee system similar to that used by the Forest Service. Subsequent amendments expanded the public land area slated for grazing districts to over 168 million acres. Each of these actions suggested the prioritization of resource conservation. And each stood as a profound declaration that the era of public land disposal was over. From this point onward, with a few exceptions,[21] the public domain would be permanently managed and protected, however weakly, by the federal government.

In contrast, in the administrative details—and even in some of the wording—the Taylor Act powerfully reflected the resource development side of its mandate, offering numerous illustrations of Senator Taylor's efforts to protect the interests of western stockmen. For example, rather than grant regulatory power to the Forest Service, the act established a new Division of Grazing within the Department of the Interior's General Lands Office.[22] Allowing cattlemen to avoid the scrutiny of a federal agency with a tradition of conservation and scientific management (e.g., the Forest Service), the new Grazing Division provided ranchers the opportunity to shape the tone and rigor of regulatory enforcement. As a result, the agency was chronically underfunded and understaffed. However, the most pronounced expression of the stockmen's influence came in the selection of the division's first director. For this post, Interior Secretary Harold Ickes appointed Farrington Carpenter, a Colorado rancher who qualified not on the basis of expertise in rangeland science but because of his ties to the western cattle industry.[23]

The industry's influence was also seen in the procedures adopted for allocating and administrating the new grazing districts. Generally speaking, boundaries were drawn through consultations between Director Carpenter, private ranchers, and state officials. Priority for receiving permits was then given to ranchers living within or adjacent to each grazing district. The former group also qualified for free domestic use or subsistence grazing permits. Those who used public rangeland prior to 1934, and those who owned enough private land and water rights to support their herd on base ranches when not on the public range, also received special preference. Likewise, stocking levels were determined according to pre-1934 AUMs. In short, under the Taylor Act, ranchers who used public rangeland in the past were likely to continue in the same manner and conditions as before.

The primary difference was that ranchers now paid a fee for the right to use public land. Set at five cents per AUM, grazing fees fell well below those charged by the Forest Service at the time (fifteen cents per AUM in 1934) and nowhere near fair market value for private pasture. Finally, to ensure even more favorable administrative control for ranchers, the

Taylor Act created local grazing advisory boards. Composed primarily of local stockmen, the Boards managed grazing district boundaries, allocated permits, determined grazing seasons, and set stocking levels. For conservation advocates, this did not sit well. It appeared that the act had resulted in a private government, a self-regulating body with little public accountability. Federal oversight without a meaningful "federal" component created a seemingly contradictory administrative situation, testing the tolerance of FDR, Secretary Ickes, and conservation activists alike. Nonetheless, the regulatory mold was now in place, setting the stage for a long series of conflicts and debates over the proper management of public rangelands for decades to come.

CREATING THE BLM

In the years following passage of the Taylor Act, public rangelands remained firmly in the hand of western ranching interests. However, in 1939, Secretary Ickes decided to shake things up. Frustrated with Director Carpenter's close relationship with stock growers and his corresponding failure to develop federal regulatory capacity within the Grazing Division, Ickes sent him packing.[24] Changing the name of the Grazing Division to the U.S. Grazing Service, Ickes turned to Richard Routledge as the new director. Under prodding from Ickes, Routledge attempted to establish conservation as a priority in the newly rechristened agency, but he found little success.

It didn't help that Congress had no interest in such a change. Through the 1940s, efforts to raise fees or reduce stocking numbers failed miserably. In 1944, Director Clarence Forsling proposed an increase in grazing fees from five to fifteen cents per AUM. He was rewarded with a round of hostile congressional hearings led by Nevada senator Pat McCarran, a powerful proponent of ranching interests. Congress not only rejected Forsling's idea but also moved to radically reduce the agency's operating budget and personnel. In 1946, Congress decided to cut the 1947 Grazing Service budget to 53 percent of 1945 levels, reducing staff by two-thirds. To make up the difference, local grazing advisory boards agreed to contribute over a third of the Grazing Service budget. But this created a highly awkward situation in which public land users now paid a portion of the salaries of their own regulators.[25]

In an effort to revive the struggling agency, in 1946 President Harry Truman issued a Reorganization Plan that merged the U.S. Grazing Service with the General Lands Office to create a new agency named the Bureau of Land Management (BLM). For livestock interests, however, this move made a good situation even more advantageous. No longer able to focus exclusively on grazing issues, the new agency had to balance myriad responsibilities, including the administration of mining claims, leases, and patents on all types of federal lands. Moreover, lacking a statutory mandate, the BLM enjoyed no more congressional support than the U.S. Grazing Service had received—a fact directly reflected in annual appropriation debates.

The summer of 1946 witnessed another effort to exert industry influence over the public domain. A conference in Salt Lake City convened livestock growers with state officials to promote a shared idea: privatizing federal rangelands along with other forms of public lands in the region, including national parks and forests. Congressional legislation sponsored by Senator Edward Robertson of Wyoming sought to transfer these lands to the western states so they could be sold off to local ranchers at reduced prices. This early "Sagebrush Rebellion" was roundly criticized by Bernard DeVoto and other environmental advocates. Writing in *Harper's Magazine*, he underscored the audacity of the proposal.

> [T]he Cattle Kingdom never did own more than a minute fraction of one percent of the range it grazed: it was national domain, it belonged to the people of the United States . . . and since the fees they pay for using public land are much smaller than those they pay for using private land, those fees are in effect one of a number of subsidies we pay them. But they always acted as if they owned the public range and act so now . . . and they are trying to take title to it. . . . The immediate objectives make this attempt one of the biggest land grabs in American history. The ultimate objectives make it incomparably the biggest.[26]

The work of DeVoto and other writers turned public opinion against these ideas. As a result, legislative efforts to privatize the public rangelands were placed on hold until the next iteration of the Sagebrush Rebellion emerged in the late 1970s. Despite these trials, during the tenure of BLM director Marion Clawson (1948–1953), some progress was made in stabilizing the agency. A Harvard-trained agricultural economist, Clawson was from Nevada and well acquainted with the livestock industry. A tireless proponent of professional administration, he established a more efficient administrative structure for the agency, restored the agency's budget, made personnel changes that helped set the tone for a more professional modus operandi, and successfully raised grazing fees from eight cents to twelve cents per AUM in support of an expanded range management program.[27]

But overall, the Eisenhower era was characterized by small strides forward, followed by powerful congressional reactions to protect stockmen's interests.[28] The political iron triangle between the livestock industry, congressional representatives, and the BLM kept significant change at bay

TEXTBOX 7.1.
A Tale of Two Logos

Changes to the BLM logo offer a unique and quite literal illustration of the agency's attempts to redefine its mission and managerial priorities over the years. Issued in 1952, the original logo or seal (figure 7.2a) featured five men: a miner, rancher, engineer, logger, and surveyor. Standing shoulder to shoulder in profile, they represent the concept of multiple use, but defined in such a way as to signify multiple forms of resource extraction and development rather than recreation or other noncommodity conservation concerns. Behind the men, the logo depicts a covered wagon train rambling across the prairie. In front of them lies a modern landscape filled with railroad tracks and an urban skyline complete with oil derricks and smoke-billowing factories. The image celebrates the march of progress fueled by the industrial transformation of raw nature.

In 1964 BLM director Charles Stoddard introduced a new logo for the agency, which is still in use today. In sharp contrast to the busy utilitarian focus of the old image, the new symbol portrays a highly idealized rendering of a conifer tree, a meandering river valley, and a distant snowcapped mountain (figure 7.2b). The new logo more closely resembles those used by the National Park Service (featuring an arrowhead, bison, and sequoia tree) and the USDA Forest Service (containing a single conifer tree).

With the men and industrial signifiers now removed, the new message stresses the conservation of resources in their natural state. While not precluding their use as commodities, the depiction opens the way—at least symbolically—for a much broader set of management priorities for BLM lands, ranging from wildlife protection and recreation to resource extraction.

Figures 7.2a and 7.2b. A comparison of BLM logos from 1952 (a) and 1964 (b) illustrates attempts to change the agency's image in terms of managerial priorities. Images courtesy of the BLM.

for most of the decade. Dominant use (grazing and mining) continued to characterize the agency much more than the concept of multiple use.[29] It was not until the 1960s that support for a stronger conservation mandate began to emerge, slowly gathering steam through the Environmental Decade of the 1970s. Led by Interior Secretary Stewart Udall, who served in both the Kennedy and Johnson administrations, efforts were made to expand recreational opportunities on all types of public lands, including those managed by the BLM.

Progress was slow at first. In 1960, Congress passed the Multiple Use and Sustained Yield Act, granting the Forest Service a clear and concise mandate for multiple-use management. But the BLM was not included in this statute, despite persistent lobbying by agency leadership. In a similar fashion, the Wilderness Act of 1964 created new wilderness areas and mandated all federal land agencies to inventory their holdings for potential future designation. All federal land agencies, that is, except the BLM. Once again, the agency stood on the outside looking in. Despite the fact that the BLM managed more land than any other public agency, it continued to lack a clear congressional mandate—an "organic act"—that could identify managerial and administrative priorities, including those relating to environmental conservation.

The first steps toward this goal occurred with passage of the Classification and Multiple Use Act of 1964. The law directed the BLM to classify all lands within its stewardship as either suitable for disposal (e.g., for sale) or retention under federal management according to a wide variety of values.[30] These values or uses ranged from traditional grazing, mining, timber, and industrial development, to fish and wildlife, watershed, outdoor recreation, and wilderness protection.[31] To the consternation of western mining and ranching interests, the agency recommended that out of 180 million acres reviewed, less than five million acres were suitable for disposal. Moreover, the agency designated portions of their domain as recreation areas and primitive areas, a prototype for future BLM wilderness. In a separate but related move, the BLM unveiled a new agency logo, sporting a design that underscored a shift in managerial focus from resource extraction to a more broadly conceived set of priorities (textbox 7.1 and figure 7.2).

That same year, Congress established a bipartisan Public Land Law Review Commission (PLLRC). Chaired by Colorado congressman Wayne Aspinall, a longtime proponent of resource development on public lands, the Commission was charged with making comprehensive management recommendations for the public domain. On June 20, 1970, the Commission issued its final report, titled *One Third of the Nation's Land*. Containing some 137 different recommendations, the report was roundly criticized from all sides. In many ways slanted toward the promotion of resource extraction and development at the expense of noncommodity priorities, it also contained some conservation-leaning recommendations. The latter included an affirmation that the bulk of public lands managed by the BLM should remain in federal control, and that grazing fees should be raised to fair market value. But perhaps the greatest impact of the report was demonstrating the need for a congressionally mandated organic act to provide guidance for the BLM.

In the meantime, while the PLLRC plugged away on its report, other events encouraged the BLM to give greater consideration to the issue of environmental protection. For example, in 1965, Secretary Udall established sixteen new Scientific Natural Areas (areas set aside for scientific research) on lands managed by the BLM. Then in 1969, he created the nation's first wild horse range on BLM lands located in the Pryor Mountains that straddle the Montana-Wyoming border. That same year, passage of the 1969 National Environmental Policy Act (NEPA) obligated the BLM to carry out environmental impact statements for all management actions.

As the Environmental Decade dawned, Congress passed a series of new laws that continued to impact the management of BLM lands. In 1971, the Wild and Free Roaming Horse and Burro Act required the agency to protect wild horse and burro herds found on the public domain and to balance their needs with those of domestic livestock growers and wildlife. A few years later, the Endangered Species Act of 1973 mandated the BLM to take steps to protect threatened and endangered species.

Of course, receiving a federal mandate and actually carrying it out can be two different things. Though by this time the BLM had adopted and actively promoted its own pro-environmental message in the guise of Johnny Horizon (see figure 7.3 and textbox 7.2), in practice, the agency had great difficulty adjusting to environmental protection as a managerial priority. In short, the BLM was ill prepared—both fiscally and administratively—to take on these new responsibilities at the level expected by the burgeoning environmental movement. But neither could it dismiss them per the request of the agency's traditional resource extraction constituencies.

These tensions were on full display in the debate over the Trans-Alaska Pipeline in the early 1970s. Mandated by NEPA to craft an environmental impact statement (EIS) for the project, the BLM was roundly criticized by environmental activists for its poor effort. Clearly coming down on the side of industry, the BLM faced lawsuits by the Wilderness Society and other environmental groups for failing to perform its legal duties. Only congressional intervention allowed the pipeline to go forward in 1973. That same year, the Natural Resources Defense Council filed suit against the BLM with regard to grazing management. Rather than complete individual EISs for each grazing allotment, the agency had chosen instead to

KEEP IT CLEAN!

U.S. DEPARTMENT OF THE INTERIOR • BUREAU OF LAND MANAGEMENT

Figures 7.3.   A Johnny Horizon poster from the "Clean Up America" campaign that lasted from 1968 to 1976. Image courtesy of the BLM.

conduct a single assessment for some 178 million acres of public range-lands. In the 1974 ruling of *NRDC v. Morton*, the court decided against the BLM, ordering it to complete 212 EISs (later reduced to 144) and to take serious account of noncommodity impacts.[32]

Amid these mounting legal pressures, the BLM found itself reeling. Although benefitting from budgetary and staffing increases designed to help the agency meet its NEPA obligations during this period, conflicts between antagonistic interest groups took a toll on the agency. Once again, BLM officials renewed calls for a congressionally authorized "organic act" to guide them through the political minefield of multiple-use management.

THE BLM ORGANIC ACT

In 1976, Congress finally granted the BLM its long-sought-after wish: an "organic act" providing statutory authority for the agency on par with that held by the National Forest Service and National Park Service.[33] The Federal Land Policy and Management Act (FLPMA) declared once and for all that lands in the public domain would remain in federal ownership, unless disposal of a specific parcel was in the national

TEXTBOX 7.2.
The Story of Johnny Horizon

In the tradition of Smokey Bear, Woodsy Owl, and even Puddles the Blue Goose, the BLM launched its own public relations campaign in 1968 with a new mascot, named Johnny Horizon. Sporting handsome "Marlborough Man" features, wearing a cowboy hat and boots, but also a backpack, Johnny spearheaded a campaign to clean up America's public lands by the nation's bicentennial in 1976.

Adopting Woody Guthrie's famous song as his motto, Johnny declared, "This land is your land . . . so keep it clean!" He encouraged Americans to volunteer for cleanup events and encouraged kids to sign pledge cards to respect the land, be careful with fire, but also "obey state game and fish laws" and "leave gates and fences" alone. Though Johnny's appearance and code of conduct strongly evoked the rural West (and the BLM), the program's success eventually led to his adoption by the entire U.S. Department of the Interior.

Program director George Gurr broadened the campaign by reaching out to the Boy Scouts of America, who received the "Johnny Horizon Award for Environmental Improvement" in 1971. But the campaign really took off when Gurr asked Burl Ives to join the cause as Johnny's celebrity spokesman. With public appearances on the *Tonight Show* and the *Johnny Cash Show*, Ives introduced Johnny Horizon to audiences across the nation. He wrote and performed Johnny Horizon songs, and he even released an entire LP record about the icon. Ives's passion brought other celebrities to the cause, including Glen Campbell and Carol Burnett, along with features in the *New York Times* and *Newsweek*. By the early 1970s, over three hundred thousand people volunteered annually in Johnny Horizon events. One could buy merchandise, ranging from T-shirts and wristwatches to coloring books, stamp sets, plant growing kits, and waste collection bags. Parker Brothers even created a Johnny Horizon Environmental Test Kit, allowing kids to conduct water, soil, and air quality tests in their local communities.

Given the campaign's success, it is a bit unclear as to why it wasn't allowed to continue beyond the original 1976 deadline. Some refer to "political obstacles" and a change in BLM leadership. Perhaps it was a case of overreach, setting up expectations for environmental preservation priorities that the agency was not yet ready to embrace. In any event, according to one commentator, "The new guys didn't like Johnny . . . said he wasn't a 'natural symbol' like a certain bear. They resented his popularity and growing budget . . . [and] was phased out." Regardless of the rationale, many who remember Johnny still view him with fondness and as a lost opportunity to change the public perception of the BLM in the midst of the Environmental Decade.

Source: J. "Mad-Logger" Lewis, "Forgotten Characters from Forest History: Johnny Horizon," in *Peeling Back the Bark*, a publication of the Forest History Society, March 17, 2011.

interest. In this way, the law effectively repealed the remaining homestead acts and most other land laws that facilitated privatization. There
were, however, a few notable exceptions. The Homestead Act continued
in force in Alaska for another ten years (until October 20, 1986), and certain foundational laws allowing for private uses on public lands, such as
the 1872 General Mining Act, 1920 Mineral Leasing Act, and 1934 Taylor
Grazing Act, remained intact.

FLPMA also directed the BLM to manage public lands in accordance
with sustained yield principles and multiple-use values, including "scientific, scenic, historical, ecological, environmental, air and atmospheric,
water resource and archeological."[34] Of course, multiple use also meant
continuing to manage for resource extraction purposes too. However, in
so doing, the act explicitly charged the BLM to take new priorities into
account: to manage in such a way as to provide food and habitat for fish,
wildlife, and domestic animals, as well as for outdoor recreation, human
occupancy, and resource development. Finally, FLPMA mandated the
protection of areas of "critical environmental concern" and granted the
BLM fifteen years to complete an inventory of lands for potential inclusion in the National Wilderness Preservation System.

Significantly, the law also ordered the BLM to demand "fair market
value" for the use of all public lands and resources under its care. Grazing
fees, however, constituted an exception. Rather than raise them, FLPMA
required a one-year freeze while Congress studied the issue.[35] Two years
later, Congress revisited grazing with the 1978 Public Rangelands Improvement Act (PRIA). The law adopted a formula favoring ranchers
by tying fees to production costs (prices of forage and beef in any given
year). At the same time, the Carter administration directed the BLM to
review stocking levels to enhance rangeland health and to initiate proposals for wilderness designations. In response, the cattle industry mobilized
congressional supporters to pass an amendment to the 1980 Appropriations Bill requiring any stock reduction orders of over 10 percent to be
phased in over a two-year period.

Despite favorable congressional actions regarding fees, stockgrowers
still felt threatened enough by the new conservation measures to launch
another Sagebrush Rebellion. In 1979, partly in response to FLPMA's
public land retention clause, PRIA's grazing fee changes, and the Carter
administration's conservation programs, western livestock interests once
again convened and called for the wholesale transfer of western public
lands to the states for eventual sale to private citizens. That same year, the
Nevada legislature passed a measure claiming authority over the roughly
forty-eight million acres of BLM land lying within state borders. Other
western states followed suit, but they soon buckled under the legal test
of constitutionality. By the early 1980s, the effort had largely fizzled out,

but not before garnering significant national attention, including from the newly inaugurated president, Ronald Reagan.[36]

1980s AND 1990s: FROM SAGEBRUSH REBELLION TO RANGELAND REFORM

Though the Sagebrush Rebellion failed, it found a friend in the Reagan administration, which worked tirelessly to roll back the regulatory and oversight powers of the federal government. For the BLM this meant reductions in budget and staffing, especially with regard to planning and enforcement programs. It also meant concerted efforts to give greater management control back to livestock, mining, and other private commercial interests. As much as possible, the administration sought to allow free market forces to guide land use decisions.

Appointing James Watt as Secretary of the Interior and Robert Burford as BLM director, Reagan charted a new course for the agency: prioritizing mineral leases, placing a hold on grazing allotment reductions, and opening wilderness study areas to resource extraction. On numerous occasions, Congress or the courts ordered Watt to reverse his actions, many of which concerned the approval of leases on sensitive public lands without due process. When Watt resigned in 1983, his replacement, William Clark, continued in a similar vein. Throughout the Reagan administration and into the George H. W. Bush administration, the BLM shifted decidedly away from the promise of a more cohesive and effective regulatory authority, returning instead to its historical role of facilitating resource development on the public domain.

In the late 1970s and early 1980s, smoldering sentiments from the 1970s Sagebrush Rebellion found new life in the burgeoning Wise Use and County Supremacy movements. Though the legal arguments of county supremacists quickly suffered the same fate as the sagebrush rebels,[37] the Wise Use Movement achieved more lasting influence. Adopting a broader message than that of the Sagebrush Rebellion, the Wise Use Movement championed the issue of property rights and federal deregulation. Consisting of a loose collection of property-rights advocates, land developers, motorized recreation enthusiasts, and representatives of various resource extraction industries, "wise users" deployed an effective populist rhetoric in pursuit of their goals. During the Reagan and Bush eras, their discontent focused on the presumed influence of environmental NGOs, though a fair share of criticism was regularly reserved for federal laws and resource managers. Favorite targets included the Endangered Species Act, wilderness area designations, and instances of "regulatory takings." But it was during the northern spotted owl controversy in the Pacific Northwest that they delivered a masterstroke: skillfully framing the debate in terms of "jobs versus the environment." Wielded to great effect in environmental conflicts up to the present day, this trope harkens back to nineteenth-century nature-as-commodity narratives that claimed only land devoid of economic value could and should be set aside for federal protection.

For BLM lands, the spotted owl debate rendered demonstrable impacts insofar as a considerable amount of BLM forestland is located within the area covered by the Northwest Forest Plan.[38] However, even greater influence was felt in the 1994 congressional elections that brought Republicans control of the House of Representatives for the first time in decades. Back in 1992, the Clinton administration came to power on the promise of government reform, including a strengthening of environmental laws. On BLM lands this translated into support for the ideals of ecosystem management, an increase in federal grazing fees to market price levels, and a legislative overhaul of the 1872 Mining Act, including the establishment of a royalty system and strict environmental conservation standards for mining practices on public lands.

President Clinton's effort to end public land subsidies to the mining, grazing, and timber industries was a crucial piece of his effort to balance the national budget. As such, many of his supporters assumed these initiatives would appeal to fiscal conservatives eager to cut government spending. Nonetheless, all attempts to enact reform through congressional legislation came to naught. Bills either drowned amid the fragmented regional politics of the Democrat-controlled House during Clinton's first two years in office or crashed like waves upon the hardened seawall of the Republican-controlled House under Speaker Newt Gingrich after the 1994 elections.

With the failure of congressional action, President Clinton's Interior Secretary, former Arizona governor Bruce Babbitt, and new BLM director Mike Dombeck, set out to achieve reform via another means: administrative rule making. To address mining issues on BLM lands, they focused on strengthening the enforcement of existing regulations. Specifically, they required mining claimants to set aside financial bonds for environmental cleanup, to craft BLM-approved mining plans for all mines (previously those under five acres were excluded), and they raised penalties for violations. Perhaps most significantly, for the first time in U.S. history, the Interior Secretary reserved the right to refuse mining claims if they posed environmental harm. Finally, in 1994, Congress came to agreement on one aspect of reform, passing a one-year moratorium on mining patents on all lands administered by the BLM.[39] Included as a rider to the Interior Department appropriations bill, Congress has renewed the moratorium every year since.[40]

Although these changes fell short of comprehensive mining reform, their effect should not be discounted. The Clinton administration's efforts essentially transformed mining from a "dominant, best and highest use" on the BLM lands to just one of many other uses. Mining could still occur (in fact, it actually increased during the Clinton years), but only if managed in such a way as to balance it with other priorities, including environmental protection.

Federal grazing fees represented an equally challenging issue for reform. Secretary Babbitt's grazing plan for BLM lands, titled *Rangeland Reform '94*, was comprehensive in scope. Reforms included: 1) hikes in federal grazing fees to better reflect current market prices, 2) an end to the practice of allowing permitees to earn new titles or water rights for improvements they made within their grazing districts,[41] 3) a replacement of the old grazing advisory boards with Resource Advisory Councils (RACs) composed of representatives from various interest groups and backgrounds, 4) a new national rangeland grazing standard based on ecosystem management principles, and 5) a requirement tying permit renewals to range stewardship.

In the end, after months of touring the West, meeting with residents, ranchers, and local politicians, and encountering staunch resistance from all sides, Secretary Babbitt settled on a final rule. It stemmed largely from collaborative meetings of the Colorado Working Group, a collection of ranchers and environmentalists working together with Secretary Babbitt and Governor Roy Romer to hammer out a compromise agreement. In the end, it contained many significant compromises, reflective of the strong regional sentiments against increased regulation of public land grazing. The final rule backpedaled entirely on the issue of grazing fees, leaving them untouched. It also dropped the requirement of a national grazing standard, allowing individual states to define them through a collaborative decision-making process. And finally, it allowed ranchers to keep existing titles and water rights linked to past land improvements, but it prohibited the granting of new rights in the future.

The new Resource Advisory Councils (RACs) stand as one of the most significant changes to emerge from Secretary Babbitt's rangeland initiative. Replacing the Taylor Act's grazing advisory boards, made up of local ranchers, the RACs mandated balanced participation by representatives from three broad groups, including 1) commercial interests (usually from the livestock, mining, or logging industries), 2) environmental or historical groups (including those with interests in wild horses and burros or "dispersed recreation," and 3) state and local government, Indian tribes, and the public at large.[42]

Comprised of twelve to fifteen members, each RAC consists of local residents selected by the Interior Secretary according to the categories

noted above, for three-year terms. They provide advice and recommendations to BLM managers on all aspects of land management within their
specific BLM region.[43] The goal is to find consensus when possible, but it
is not required. The forums provide a formal, institutionally supported
venue for pursuing collaborative discussion and learning among diverse
interests. In this respect, proponents view RACs as one of the best hopes
of moving the management of BLM lands toward an ecosystem-based,
sustainable management framework.

GRAND STAIRCASE-ESCALANTE NATIONAL MONUMENT

In addition to mining and grazing reforms, the Clinton Era ushered
in several additional changes that served to further the conservation
mandate for BLM lands. As the 1996 presidential elections approached,
members of President Clinton's reelection campaign proposed actions to
shore up support from the environmental community. Given the current
stalemate in Congress, one idea was to use the Antiquities Act to declare
a new national monument. Prepared behind closed doors, the plan was to
issue an executive order just prior to the election, thereby giving the president an electoral boost without the burden of prolonged political debate.
    Interior officials quickly settled on a 1.7-million acre[44] section of BLM
land in southern Utah (figure 7.4). Situated near other national parks and
monuments, the remote landscape of deserts and canyons provided an
ideal opportunity to protect environmental resources and perhaps re-
direct the mission of the BLM. For environmental advocates, the timing
could not have been better. Currently embroiled in debates over proposed
coal mining leases and wilderness area designations in the region, activists needed federal help to prevail. Yet like most local residents, the entire
Utah congressional delegation, and even BLM staff, they had little fore-
warning of the president's intentions. The small staff appointed by Secretary Babbitt kept the plan secret until the day of the announcement. On
September 18, 1996, in a ceremony held on the rim of the Grand Canyon,
President Clinton signed the declaration creating the Grand Staircase-
Escalante National Monument.
    Inasmuch as the declaration received cheers from environmental
groups, it triggered vehement reactions from Utah politicians, ranchers, and the mining industry. Utah senator Orrin Hatch called it the
"mother of all land grabs," likening the president's secrecy to the attack
on Pearl Harbor.[45] Utah governor Mike Leavitt declared the action "one
of the greatest abuses of executive power in history."[46] Local residents
and county officials, hoping to benefit from pending mining projects,
expressed their displeasure with demonstrations that included burning

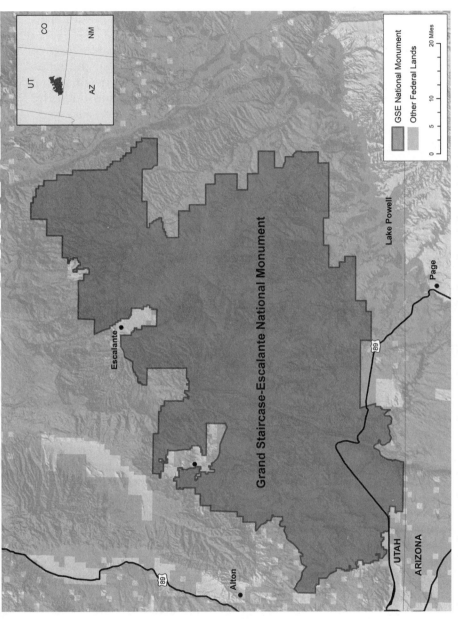

Figure 7.4. The Grand Staircase-Escalante National Monument in Utah. Map by Jessica Lee.

effigies of President Clinton and releasing black balloons as a warning to other states of what he might do to them next.[47]

Despite the political uproar, perhaps the most significant aspect of the Grand Staircase declaration was the decision to keep the lands under BLM stewardship. Up to this time, the creation of new national monuments from BLM land meant the transfer of managerial authority to other agencies, most notably to the National Park Service, such decisions implied that the BLM was simply not equipped—neither in terms of expertise, resources, nor institutional will—to properly manage lands in which preservation and recreation activities were paramount. The Grand Staircase declaration thus served as a major statement of confidence from the Clinton administration that the BLM was up to the task, ready and able to move with intentionality toward a true multiple-use mandate. Coupled with the 1994 California Desert Lands Act, which placed hundreds of thousands of acres of wilderness and wilderness study areas under BLM management, the agency now held an unprecedented conservation responsibility.

THE NATIONAL LANDSCAPE CONSERVATION SYSTEM

In the wake of the Grand Staircase declaration, Secretary Babbitt worked to identify additional BLM lands for national monument status. With President Clinton eager to leave a conservation legacy, Babbitt enjoyed full support in this quest. But this time around, he opted for a more open process, engaging local residents and BLM field officers in public discussions. In addition, he focused on less controversial locations, places already in frequent use as sites for recreation or previously designated as scientific natural areas or wilderness study areas. Ultimately, Clinton established fourteen additional national monuments by the end of his second term, totaling nearly five million acres of BLM lands.

To ensure the longevity of the monuments within the BLM management structure, Babbitt established individual appropriations and full-time staffing for each monument. Then, in June 2000, he introduced the National Landscape Conservation System. Rather than a new type of land classification, the NLCS collected together all of the existing specially designated lands within the BLM system. These included the new national monuments, along with wilderness areas, wilderness study areas, national conservation areas, recreation areas, wild and scenic rivers, national scenic and historic trails, and the desert conservation lands in California.[48]

Babbitt intended the NLCS to assist in the transformation of the BLM into an agency with a true multiple-use conservation focus, both externally, in terms of the agency's public persona, and internally, within the culture and values of the institution. Writing in 2003, Babbitt explained,

> My hope is that, by endowing the BLM with a high profile conservation mission, the old bureaucratic mule will awaken to a new future as environmental steward right up there with the National Park Service and the National Wildlife Refuge System. The day is coming, I believe, when the BLM, so often stereotyped and dismissed as the Bureau of Livestock and Mining, will be better known as the Bureau of Landscapes and Monuments.[49]

Currently, the system includes 887 areas, comprising approximately twenty-seven million acres or just over 10 percent of all BLM lands. However, several commentators note that the NLCS has yet to live up to former secretary Babbitt's expectations. This is due in part to the influence of the George W. Bush administration, which inherited the new system and assumed responsibility for its initial implementation. Unlike the Grand Staircase Monument, the new monuments received funding and staffing according to Bush administration priorities. Not surprisingly, they failed to receive levels of support proportional to Grand Staircase-Escalante. For the first eight years of the new millennium, Secretary Babbitt's vision for a new BLM and NLCS was placed on hold.

2000s TO THE PRESENT: TO DRILL OR NOT TO DRILL?

Upon taking office, the G. W. Bush administration moved quickly to reassert resource development as the primary focus for BLM lands. Bush's choice for Secretary of Interior was Gale Norton, former attorney general of Colorado and protégée of former interior secretary James Watt. And for BLM director, he selected Kathleen Clarke, former head of Utah's Department of Natural Resources and a known proponent of mineral development on public lands.

Like President Clinton before him, President Bush soon found Congress to present a less than satisfactory path to reform. Legislative victories were few, but there were notable exceptions. The 2003 Healthy Forest Act opened the door to expedited logging on national forests and BLM lands, with the rationale of thinning dense stands in order to reduce fuel loads and thereby lessen the risk of massive forest fires. The 2005 Energy Bill represented another Bush victory, opening the way for increased mineral development on BLM lands. In the wake of 9/11 and dual wars in Afghanistan and Iraq, arguments for increased energy independence as a national security measure carried significant political weight. However, rather than focus on conservation and alternative energy development, the Bush White House, under the guidance of Vice President Dick

Cheney, aggressively promoted increased fossil fuel production on the public domain. The so-called Halliburton Loophole exempted a new process called hydraulic fracturing from the Safe Water Drinking Act, thereby facilitating the rapid expansion of national gas drilling on Marcellus Shale formations on federal and private lands across the nation. Nevertheless, amendments to open ANWR to oil drilling were left out of the final legislation.

Despite these legislative accomplishments, President Bush increasingly looked to executive orders and administrative rule changes as the most effective means of advancing his agenda. Learning from the miscues of the Reagan administration and that of his father, Bush approached the task in a different way. Using nuanced language that situated his initiatives within a "pro-environment" rhetoric, the administration's deregulatory efforts rarely caused alarm amid the general public. Rule changes were even announced in a strategic way, often with press releases given late on Friday afternoons, a time when many people were already given over to weekend plans and no longer attuned to the events of the day. In this way, President Bush placed a hold on many of President Clinton's new environmental rules and declarations, slating them for a second round of administrative review prior to implementation. Such was the fate of the Roadless Area Rule, the new mining standards, and, of course, President Clinton's newly declared national monuments and wilderness study areas.

In 2001, President Bush withdrew the ability of the Interior Secretary to reject mining claims based on environmental impacts. And in 2003, he moved to undo key components of Secretary Babbitt's rangeland reform initiatives. These included: 1) returning partial ownership of rangeland improvements to ranchers (titles and rights), 2) ordering that any reductions in stocking levels on grazing permits be phased in gradually over five years, 3) limiting opportunities for public comment on grazing management plans, and 4) changing regulatory definitions to make it more difficult—and therefore, unlikely—for BLM managers to declare rangelands "unhealthy."[50] These grazing policy changes were short-lived, however, as most were thrown out by federal courts in 2007. After years of legal wrangling, the same fate befell President Bush's efforts to rescind the Roadless Area Rule in 2009 (chapter 5).

Mining, however, was a different story. The Bush administration effectively steered the BLM back toward traditional priorities, maintaining the status quo with regard to public lands grazing and overseeing a vigorous upsurge in mineral leasing across the West. A number of these leases were located on what many deemed to be sensitive lands near Arches and Canyonlands National Parks in southern Utah, and Dinosaur National

Monument in the northeastern part of the state.[51] The robust conservation mandate envisioned by Secretary Babbitt appeared all but lost.

However, with the arrival of the Obama administration in 2009, the pendulum began to swing back once more. As one of his first actions as Secretary of the Interior, Ken Salazar suspended the mining leases granted near the desert parks and monuments in Utah pending further study of their environmental impacts. Passage of the 2009 Omnibus Public Lands Management Act produced significant effects as well, setting aside 1.2 million acres of the Wyoming Range from oil and gas development, expanding wilderness designations within BLM land, and most importantly, breathing new life into the BLM's National Landscape Conservation System by granting it formal congressional recognition.[52]

The question is, How far in the opposite direction will the pendulum actually go? It is safe to say that the BLM is renewing commitments to a multiple-use approach whereby land preservation, environmental health, and recreational opportunities are once again given greater consideration in overall management planning and implementation. But under the Obama administration, these activities still remain dwarfed beside traditional resource extraction management responsibilities. For every move toward environmental preservation on BLM lands, there come additional initiatives to expand resource development. For example, following the 2009 mining lease suspensions in Utah and the permanent protections established in the Wyoming Range, Secretary Salazar announced in 2011 an expansion of new coal leases encompassing nearly 7,500 acres in Wyoming's Powder River Basin.[53] Similarly, in 2012, as part of the latest reversal of Bush Era mineral leases, Secretary Salazar signed a twenty-year ban prohibiting uranium and all other hard rock mining within a one-million-acre buffer zone surrounding the Grand Canyon. But significantly, the ban did not preclude new oil and gas, geothermal, or other types of mineral leasing, nor did it impact the nearly 3,200 current mining claims located within the zone. In response to claims by Republican opponents that this decision will nonetheless serve to cripple the regional economy, Salazar pointed out that over twenty-five million people depend on the watershed for drinking water and irrigation.[54] Moreover, he argued that the recreation and tourism economy built around a healthy Grand Canyon brings in $3.5 billion per year, dwarfing the jobs and income produced by the mining industry.

But despite this rigorous defense of environmental protection programs, the production of oil and natural gas has increased under President Obama, not only on private lands (dominated by the Marcellus Shale boom), but on federal lands as well. This has occurred despite the fact that the United States suffered its worst oil spill in history in the Gulf of Mexico in 2010, and that the number of permits and leases granted on federal lands actually decreased by 37.4 percent and 42.4 percent, respectively, during Obama's first term.

Environmentalists see a similarly mixed record on questions of livestock grazing and support for the National Conservation Lands System. They remained concerned that the administration is simply too willing to accept the status quo rather than break new ground toward a more robust conservation mandate for the BLM. Hence, the question remains: Is the BLM ready to make new strides toward noncommodity management priorities? Or will the weight of its resource extraction history—and the legacy of nature-as-commodity thinking—limit any such possible futures?

CASES

One constant in the preceding account of BLM lands is the dominance of the idea of nature-as-commodity. Secretary Salazar's response to his Republican critics over Grand Canyon protections is a case in point. As noted in the introduction to this chapter, for BLM lands, the glare of commodity thinking is often so strong, so blinding, that it can wash out all other considerations of value in nature. Throughout most of the history of BLM lands, notions of inherent, ecological, or aesthetic worth rarely enter conversations about management priorities, unless first filtered through the calculus of economic gain. But must it be this way? Is there another way forward?

The three case studies below explore the legacy of commodity thinking on BLM lands with regard to public land grazing. The first case explores the question of fees: Should they be raised to fair market value? Focusing on arguments pro and con, the case highlights the Obama administration's recent response. The second case revisits the Grand Staircase-Escalante National Monument. After a contentious beginning, how has the monument fared in its pursuit of conservation priorities? Grazing has played a central role in these efforts via the concept of "conservation use permits," reflecting both the BLM's institutional constraints and possibilities. In the final case, the Malpai Borderlands Group casts public land grazing in yet another light, as local ranchers, environmentalists, and federal officials collaborate to develop more sustainable land use practices outside formal frameworks, working across the boundaries of public and private space.

Grazing Fees: Déjà vu All Over Again

The issue of grazing fees runs like a thread through the historical evolution of the BLM. It is reflective of both the rationale for the creation of

BLM lands as well as one of the most pronounced reasons for the BLM's inability to manage them according to a robust conservation mandate. As we have seen, from the travesty of the Big Die-Up in the 1880s to the Dust Bowl conditions of the 1930s, the nation's rangelands have suffered significant degradation over the years from overgrazing. Such crises led directly to the 1934 Taylor Grazing Act and the eventual creation of the Bureau of Land Management in 1946. However, the BLM's regulatory framework has always lacked the political power to effect meaningful change in grazing practices. From the beginning, the commanding political influence of the livestock industry forged a classic iron triangle between stockmen, western congressional representatives, and the BLM itself.

Currently, the BLM administers some eighteen thousand grazing permits on public lands.[55] But aside from the issue of stocking levels, much of the political debate has centered on the question of fees. While incremental increases have occurred over the decades, all efforts to bring them up to contemporary market prices have failed. By and large the debate pits environmentalists, federal range managers, and fiscal conservatives against property rights groups, the livestock industry (both cattle and sheep), and many western politicians. Their respective arguments, in abbreviated form, are these:

Environmentalists begin with the assertion that domestic cattle and sheep constitute invasive species. Moreover, the grasses they find most palatable are not necessarily adapted to withstand the onslaught. As forage begins to diminish, livestock become less selective, at times removing the entire plant stem rather than just the uppermost leaves and stocks. This leads to soil erosion and impedes the plant's ability to reproduce. The resultant denuded conditions open the way for "early seral species": plants that thrive and grow quickly in open, bare mineral soil. Often these new plants are invasive weed species, which soon replace native grasses.

A second line of argument focuses upon soil compaction. When ranges are overcrowded, cattle hooves can compress the soil, making it difficult for new seeds to germinate or perennial grasses to reemerge. Such conditions can also prohibit groundwater absorption, thereby reducing aquifer replenishment and exacerbating rates of runoff and flooding after storm events. In addition, livestock can deplete limited water resources simply through consumption. Even with ranchers' improvements that redistribute water supplies, under drought conditions competition with wildlife may still render negative effects.

Impacts to riparian areas—the buffer zones surrounding streams and rivers—receive special attention in environmentalist critiques. Cattle often prefer to spend extended time in such places, but in the process they can wreak tremendous ecological havoc. Eating away vegetative cover, including young tree saplings such as willows and aspen, produces severe erosion. And as livestock defecate in waterways and break down stream banks with their hooves, waters may fill with sediment. This can lead to changes in stream quality, depth, and temperature that impact not only terrestrial wildlife but also a wide array of aquatic species.

For environmental activists, the ranchers' so-called improvements can also be detrimental to wildlife. Fences can block off migratory routes or access to water sources, and barbed wire can cause bodily harm, especially in crisis events such as wildfire. And if ranchers provide supplemental feed to livestock on the public rangelands, it can contain seeds that spread invasive and noxious weed species. Once established, such species can never be fully eradicated but only slowed with the use of herbicides.

Environmentalists also point to the historical impacts of federal predator control programs that served to extirpate species such as grey wolves and grizzly bears on behalf of stockgrowers. Or they highlight the negative impacts to the recreational and aesthetic values held by hikers and hunters, many of whom may find their "wilderness experiences" degraded by the presence of domestic livestock. Finally, they may refer to studies by the U.S. General Accounting Office (GAO)[56] that underscore the historically poor conditions of rangelands through the decades, and how those who benefit most from public land grazing are not necessarily small-scale family ranchers—the iconic cowboy images that so powerfully capture the public imagination—but rather large corporate interests.[57] For all of these reasons, activists argue that the current fee structure and management regime amounts to little more than a subsidy to the western livestock industry—what authors George Wuerthner and Mollie Matteson call "welfare ranching."[58]

Those who graze livestock on public ranges, of course, take a dramatically different view. With regard to grazing permits, the first point is that below-market-level fees are indeed fair and appropriate. They offer several reasons. First, public rangelands tend to be of lower quality and maintain less forage than privately leased pastures. Second, grazing permit holders must bear the costs of rangeland improvements themselves, including building and maintaining fences, gates, and cattle guards; the construction of water features such as ponds, springs, and small reservoirs; as well as efforts to protect riparian areas. These items are generally included in the price of leasing private pasture, but on public lands, the rancher must bear these costs alone.

In addition, ranchers assert that livestock grazing has a net positive ecological effect on public lands. Certain improvements can provide ecosystem services. In some cases, livestock watering ponds constitute new wetlands that support entire microecosystems. These areas can enhance biodiversity and provide water for myriad wildlife species that benefit recreationalists and hunters. The fact that they frequently ride the range to check fences and move livestock provides an additional set of eyes and ears that can assist BLM and other federal land managers. They can monitor the effects of management projects and even the impacts of other users—recreational, timber, or mining—bringing potential problems to the attention of federal officials who lack resources for regular patrols. Finally, ranchers point out that their privately owned pastures, many of which border public lands, provide critical forage for wildlife during the winter. Keeping ranching operations viable with reduced grazing fees provides an important wildlife service that benefits all users of the federal land system. If forced to sell their ranches, these open pastures may be transformed into subdivisions, leading to the loss of critical wildlife winter habitat. Such outcomes would not only harm biodiversity but also the "wilderness experiences" sought by hunters, recreationalists, and other public land users.

For all of these reasons, ranchers submit that charging market rates for public grazing permits is a miscalculation that ignores true costs and benefits. It also explains why ranchers wish to retain title to the improvements they provide, whether this amounts to water rights or equipment such as fencing. They argue that public land grazing is important not only for ranchers' own livelihoods, but also for the economic well-being of rural communities, including businesses linked to recreation and environmental tourism. For many public land ranchers, the permits have been held for generations and are fully integrated into the value of their ranches. Since many operations are also economically marginal, any drastic changes in allowable AUMs or fee hikes could render dramatic effects to their viability, negatively impacting their families and local economies.

Despite these strong differences of opinion, the status quo has remained remarkably steady over the years. Though BLM grazing fees have fluctuated from five cents to $2.36 per AUM since 1940, they have remained at $1.35 per AUM for the majority of years since 1996. This is the same fee that was charged in 1985. Over the past fifteen years, attempts to raise grazing fees, either through congressional action or administrative decree, have been met with stiff political resistance by western livestock interests. What is truly amazing about this influence is how small the industry actually is (at least in terms of numbers employed) in comparison to its political force.[59] As of 2012, the grazing fee on BLM lands remained at $1.35 per AUM.[60]

If one takes a narrow economic view of the matter, things do not look too good for public land grazers. A 2005 report by the General Accounting Office found that during the period 1980 to 2004, while the cost of private grazing pasture increased by 78 percent, the fees charged by the BLM and Forest Service during that same period actually fell 40 percent.

Moreover, in fiscal year 2004, the cost of all federal grazing programs to taxpayers was $115 million.[61] According to the GAO, in order to simply recover program costs, BLM fees would have to be raised to $7.64 per AUM.[62] Ironically, a stronger case can be made in favor of the livestock industry if one moves away from a narrow nature-as-commodity view and considers the ecological and broader indirect benefits of maintaining a place for ranching on the public rangelands.

More recently, in January 2011, BLM director Bob Abbey and Deputy Forest Service chief Joel Holtrop responded negatively to a petition by a number of environmental groups requesting an increase in grazing fees.[63] The Obama administration, via the BLM director's letter, pointed to the fact that agency resources were already stretched thin dealing with an array of mining issues on the public domain. But any thoughts that this decision was not a true reflection of the administration's position were dashed eleven months later. In November, the Public Employees for Environmental Responsibility (PEER) filed a complaint against the BLM regarding its decision to exclude grazing impacts from planned compre-hensive ecoregional assessments of western landscapes. For PEER, this move exposed the BLM's willingness to allow political influence—in this case, the livestock industry—to shape what is meant to be a scientific pro-cess. While the BLM denies this charge, some environmental groups see the legacy of commodity thinking to be as strong as ever at the agency. Raise grazing fees? For some activists, the odds of reversing climate change are much better. But for other environmental advocates, this is the wrong fight altogether. They see how the ecological benefits of ranching, depending on the way it is managed, can outweigh the costs. For these en-vironmentalists, it is far preferable to work together with livestock grow-ers through collaborative partnerships, such as the Malpai Borderlands Group, in order to achieve shared conservation goals (discussed below).

Revisiting the Grand Staircase

Returning to the Grand Staircase-Escalante National Monument, one might ask, what has been its fate since the 1996 declaration? As the sym-bol of a new day for the BLM, the Grand Staircase stands as a sign of trust from an environmentally minded presidential administration that the agency could grow and mature beyond the "captured" interest of resource extraction industries. But did the BLM succeed? The answer to this question sheds light not only on the future of the BLM but also offers another perspective on the grazing issue.

The first strike against the new monument involved bulldozers. A few days after the dedication ceremony, county commissioners from Garfield, Kane, and San Juan counties authorized bulldozing to cut or "restore" roads through wilderness study areas within monument boundaries. The goal was to discredit their qualification for wilderness protection and assert local rights of way under an obscure section of an 1866 mining law.[64] In response, the federal government filed lawsuits against all three counties.

Several weeks later, monument opponents filed a suit of their own. The Mountain States Legal Foundation (the firm founded by former interior secretary James Watt), the Utah Association of Counties, the Utah Schools and Institutional Trust Lands Administration, and the State of Utah joined forces in legal action to repeal President Clinton's declaration.[65] The litigation dragged on for eight years. But finally, in a 2004 ruling reminiscent of the federal court decision protecting Teddy Roosevelt's 1908 creation of the Grand Canyon National Monument, President Clinton's actions were upheld as a legal use of the Antiquities Act.

With the monument's future secured, mining projects blocked, and road building intrusions on hold, the next big issue involved livestock grazing. In many ways, the Grand Staircase-Escalante region of southern Utah was (and remains) cattle country. In fact, at the time of declaration, much of the Grand Staircase-Escalante National Monument was permitted for grazing. Careful not to risk the ire of ranching interests, the Clinton administration included stipulations protecting current grazing permits within monument boundaries. It also ordered the BLM to create a management plan within the next three years to balance continued grazing and limited mining activities with new recreation and preservation priorities. The new plan did just that: emphasizing scientific research (archeology, paleontology, desert ecology) and recreation opportunities within a "primitive frontier" setting—meaning no new paved roads, parking lots, or national park–style visitor centers.[66]

To assist in the development of these other uses, the Grand Staircase-Escalante Partnership (GSEP) was formed. Numbering some two hundred local residents and supported by the Sonoran Institute in Arizona, the Partnership manages the monument's volunteer programs, staffs the visitor centers, and coordinates community education and outreach programs. These include local events and support for new businesses that build upon the recreation, wilderness, and scientific management priorities in the monument. To date, the Partnership has established an annual arts festival, developed a research facility for visiting scientists, and is currently building a museum dedicated to local cultural heritage. Interestingly, a recent brochure promoting tourism and visitation to the monument for scientific expeditions, wilderness experiences, and restoration projects highlights the idea of "wilderness as a commodity." Perhaps this is an attempt to appeal to local residents who do not readily see the economic benefits of nonresource extraction uses. Nevertheless, the

phrase resonates strongly with the legacy of nature-as-commodity thinking that has so long underscored BLM lands.

In addition to the GSEP, to support the agency's new recreational and
scientific management priorities at the Grand Staircase, the BLM began
working with other regional groups, including the Grand Canyon Trust
(GCT). An environmental organization with extensive experience in ar-
ranging conservation easements with private landowners, the GCT is also
unique for its development of a new kind of easement program designed
specifically for public rangelands called "conservation use permits."[67]

First proposed as part of Secretary Babbitt's Rangeland Reform, such permits allow individuals to acquire public grazing permits and then volun-
tarily retire them by simply choosing not to graze any animals on the land
and/or requesting changes in BLM management plans to terminate the
permits altogether. The first attempts to establish conservation use permits
were met with a legal challenge brought by ranching interests. The case
went all the way to the Supreme Court. Ultimately, the court ruled in favor
of the stockmen, invalidating the idea that grazing permits could be used
exclusively for conservation purposes. Rather, they must be put to their
intended use: livestock grazing. However, the court upheld the notion that
anyone, even nonranchers, could own a permit. And it affirmed the BLM's
authority to manage permit use, including the possibility of suspension.

Nonetheless, the vague and open-ended nature of the BLM's stewardship mandate under FLPMA,[68] together with the conservation language
in the Grand Staircase-Escalante declaration, allowed federal managers
some flexibility on the permit issue. As a consequence, the GCT continued to successfully negotiate allotment buyouts with willing grazing
permit holders, mostly ranchers who wished to move elsewhere to avoid
conflicts with recreational users on monument lands. However, to stay
within the stipulated legal bounds, the GCT got into the ranching business, setting up the Canyonlands Grazing Corporation. Ironically, an
environmental NGO seeking to stop livestock grazing on public lands
now found itself in the position of owning a livestock corporation that did
just the opposite. The environmentalists were now ranchers! However, in
practice, the GCT grazed only minimal numbers of cattle on their permits
when necessary. Meanwhile, they continued to work with BLM managers
to retire their permit allotments in future management plans.

But managerial changes such as these did not come without growing
pains. In the drought-filled summer of 2000, monument supervisor Kate
Cannon asked several permitees to remove their livestock from public
rangelands early in the season to allow forage to recover. When the ranchers refused, Cannon ordered the cattle rounded up and sent to public
auction. The ranchers responded by driving to the auction barn and, with
support of the local sheriff, taking back their cattle.

Incensed by this expression of BLM's regulatory power and encouraged by the Bush administration's views toward public lands, the Canyon Counties Rural Alliance—a collection of conservative local officials, ranchers, and property rights activists—mustered its lobbying power in Utah and Washington, D.C., in a campaign to have Cannon removed. In 2003, she departed for a post at Grand Canyon National Park. With victory achieved, the Rural Alliance, in conjunction with Kane and Garfield counties, next set its sights on the issue of conservation use permits. To this end, the group filed an administrative complaint with the BLM against the Grand Canyon Trust.

However, in 2006, a BLM administrative law judge ruled in favor of the GCT. Since the organization owned a fully functioning cattle operation, complete with base ranchland and, of course, grazing cattle, the GCT was in full compliance with the law. The final fate of the GCT's permits will not be settled until the next grazing management plan for the Grand Staircase-Escalante National Monument is issued. Meanwhile, the conservative Rural Alliance continues the fight over road building and wilderness area designations in the monument.

The Malpai Borderlands Group

Another theme running through the history of BLM lands is the notion of local management. From the Taylor Act's grazing advisory boards to Secretary Babbitt's Resource Advisory Councils, in one form or another, local groups have enjoyed significant influence in shaping policy and management decisions on BLM lands. As one of the few ideas to receive enthusiastic support from both Democratic and Republican administrations,[69] the decentralized nature of BLM management has been a constant defining characteristic since the beginning.

Since the mid-1990s, the RACs have provided an official forum for collaborative discussions of public land management initiated and convened by the federal government. But emerging in parallel with the RACs, there have been other types of collaborative efforts pertinent to BLM lands. In some instances, these more informal, grassroots-initiated collaborative groups have led the way in the development of alternative approaches to sustainable resource management on public lands.

The Malpai Borderlands Group (MBG) is one of these. Standing as a quintessential example of community-based collaboration, the MBG seeks to sustainably manage a pyramid-shaped eight-hundred-thousand-acre area covering portions of New Mexico and Arizona down to the U.S.-Mexico border (figure 7.5).[70] The group takes its name from the Malpai Ranch, which in turn derives its title from the volcanic malpai rock characteristic of the region. Approximately 53 percent of the area is privately

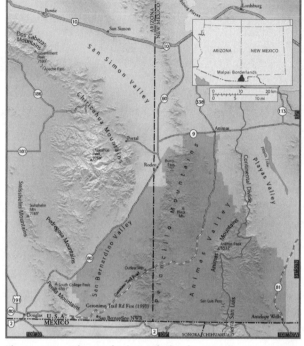

Figure 7.5. The Malpai Borderlands stretch across state boundaries in southern Arizona and New Mexico. They consist of a mix of private property and lands managed by the BLM, Forest Service, and USFWS. Map courtesy of the Malpai Borderlands Group.

owned, while 47 percent consists of public lands managed by the BLM, Forest Service, and USFWS.[71] The MBG began with a meeting of local ranchers in the early 1990s over concerns with declining range conditions and the potential for increased federal regulatory restrictions emanating from the pressure by environmental groups. Their greatest fear, however, was the threat of fragmentation as neighboring ranchers sold off their land for twenty-acre "ranchette" subdivisions.

However, rather than take a confrontational stance on these issues, falling back on litigation or congressional lobbying to force their agenda, the ranchers opted instead to reach out to local residents, including environmentalists, range scientists, and federal managers. For BLM and Forest Service officials, the prospect of working with—instead of against—local residents toward shared land management goals was far superior to the confrontations of the past. Given thinning budgets and staffing levels, local support was essential for properly implementing and monitoring conservation and restoration projects.

Meanwhile, a number of local environmental activists and ecological scientists (in many cases, the same person) were rethinking their own views about ranching. For a growing number of reasons, the idea of maintaining a "working landscape" with large chunks of land devoted to livestock production looked increasingly favorable to a future of subdivision development. First of all, keeping lands as pasture provided grassland habitat for wildlife as well as for cattle. They played an important role, offering a critical low-elevation habitat for migrating species that summered in the high-elevation public lands (often national forests) but returned to river valleys and plains in the winter. Second, it was much easier for environmental groups to negotiate effective conservation easements with a handful of large landowners than to broker hundreds if not thousands of such agreements with a myriad of suburban or ex-urban residents. Moreover, subdivision development rendered a host of negative environmental impacts, ranging from fragmented habitat to pollution runoff. In short, for many former opponents of public land grazing, if the loss of permits meant the loss of private ranches to subdivisions, then they might prefer "to see a cow rather than a condo" after all.

According to Executive Director Bill McDonald, two events spurred the formal creation of the Malpai Borderlands Group. The first concerned an outbreak of a small fire in an area of mixed public and private lands. Ecologists and ranchers had come to realize that rangeland deterioration—or rather the loss of native forage to woody plants—was due in part to federal fire suppression policies. Restoring natural fire regimes (or fire ecologies) might allow grasses to replenish themselves. However, in this case, the insistence by BLM managers to suppress the fire against the protests of local landowners brought the issue to a head. The second event concerned the sale of one of the largest ranches in the area. The Gray Ranch, spanning more than 502 square miles, was bought by The Nature Conservancy (TNC). Fearing that TNC would donate the land to the BLM or Forest Service, one local rancher agreed to create an NGO, the Animas Foundation, to buy the ranch from TNC under the condition that conservation easements be maintained.

In 1994, the Malpai Borderlands Group was officially established as a 501(c) (3) nonprofit organization. Working with federal managers, local ranchers, environmentalists, scientists, and residents in the region, the first project involved a comprehensive fire plan designed to reintroduce the natural fire regime with a program of prescribed burnings. But perhaps their most significant contribution to sustainable grazing management was the invention of "grass banking." A grass bank works by making grass available to local ranchers in times of drought or flood in exchange for conveyance of land use easements prohibiting subdivision development. Given the uneven effects of drought and precipitation patterns that can leave some pastures underwater while others remain bone dry, ranchers needed access to different pastures at different times. The idea harkens back to John Wesley Powell's recommendation for a regional grazing commons.

The Malpai Group uses pasture from the Gray Ranch (now owned by the Animas Foundation) to provide grass bank forage to local stockmen who wish to participate. All easements were (and remain) voluntary arrangements, drawn up according to the unique conditions of specific ranches.[72] In its first ten years, the MBG completed twelve new easements, covering seventy-seven thousand acres of land in the region. Four of these were tied to the use of the Gray Ranch grass bank. Counting the conserved lands of the Gray Ranch, this amounts to over half of the private land in the region now protected from subdivision development.[73]

In addition, the Group has coordinated a number of prescribed burns, including the largest in U.S. history. It negotiated a Safe Harbor Agreement and Habitat Conservation Plan with the USFWS to protect endangered species in the area, including the jaguar, leopard frog, long-nosed bat, and ridge-nosed rattlesnake. In conjunction with federal agencies, the MGB also developed technical and cost-share programs for local landowners to conduct conservation projects on their land; established a watershed restoration program; and created an outreach initiative to ranchers in Northern Mexico. Increasingly, group members offer workshops throughout the United States on their land management programs.

According to its mission statement, the purpose of the Malpai Borderlands Group is

> to restore and maintain the natural processes that create and protect a healthy unfragmented landscape to support a diverse, flourishing community of human, plant, and animal life in our borderlands region.[74]

However, as Bill McDonald explains, the MBG's greatest legacy is demonstrating how to cultivate effective and supportive working relationships with federal managers and environmental groups, without falling into the old "Sagebrush Rebellion" pattern of confrontation, litigation, and congressional lobbying. A central concept promoted by the MBG is the notion of the "radical center": the idea that compromise represents an audacious but critical approach to problem solving in the age of zero-sum politics. By embracing this ideal, the MBG continues to work collaboratively toward common goals based on the sustainable coexistence between diverse human communities and wild nature. Perhaps this approach offers the best chance of breaking the continuity of narrowly envisioned nature-as-commodity thinking that has so long plagued the lands of the BLM.

8

National Wilderness Preservation System

Wild and Scenic Rivers and National Scenic Trails

We simply need that wild country available to us, even if we never do more than drive to its edge and look in. For it can be a means of reassuring ourselves of our sanity as creatures, a part of the geography of hope.

—Wallace Stegner[1]

The clearest way into the Universe is through a forest wilderness.

—John Muir[2]

Pine needles crunch softly underfoot, and dappled sunlight filters through the forest canopy, lighting up moss-covered logs and the yellow-red bark of ancient conifers. Golden leaves of quaking aspen shimmer in the cool mountain breeze, held fast in clusters of pearled, snow-white stalks. Gaps in the trees reveal distant snow-covered peaks, jagged and sharp against a deep blue sky. Perhaps this is what John Muir was on about. Creeks babble, chipmunks squabble, hawks screech, and all the while cottony wisps of cloud sail overhead. But one thing is conspicuously absent: the imprint of human beings.

OK. I'm here, so that's not entirely true. What is actually missing, and significantly so, is blatant evidence of modern industrial society on the landscape. No roads, no parking lots, no buildings or permanent structures can be seen. No motorized vehicles are present, nor even the humming buzz of mountain bikes whizzing by. No trees slated for logging, no lands drilled, scraped, or excavated in search of minerals or fossil fuels. For this is Wilderness, the most highly protected form of federal lands in the United States.[3]

But what exactly *is* wilderness in the context of public lands? Where did it come from? And how wild is it, really? As a designated unit in the federal land system, a wilderness area is defined by Congress according to the 1964 Wilderness Act (see textbox 8.1). By law it is a relatively large area (five thousand acres, or of "manageable size") where humans visit but do not remain, where the imprints and alterations of society are but temporary, and where resource development is severely limited or prohibited altogether. But does this legal definition capture the essence of wilderness and why it matters so deeply to so many people?

Another way to define it is in terms of managerial priorities. If BLM lands lie on one end of the management spectrum, wilderness areas occupy the opposite position. They complete the continuum of federal land priorities and values: from resource development through conservation to preservation. However, unlike BLM lands, national wildlife refuges, and most other types of federal land units, there is no single federal agency with exclusive managerial authority for wilderness.[4] Created through acts of Congress, each wilderness area is assigned to one of the four federal land agencies (National Park Service, Forest Service, U.S. Fish and Wildlife Service, or Bureau of Land Management) as part of the National Wilderness Preservation System. The Forest Service thus manages wilderness areas located within national forests, the Park Service oversees wilderness in national parks, and so on.

As of 2013, there were 757 wilderness areas, covering approximately 109.5 million acres (figure 8.1).[5] The Park Service manages the largest amount, some 40 percent of the total over sixty units. The Forest Service accounts for about 33 percent on 431 units, though many wilderness areas managed by other agencies today were once part of the national forest system. Meanwhile, the national wildlife refuge system and BLM lands make up the remaining 19 percent and 8 percent, respectively.[6] Perhaps not surprisingly, the BLM wasn't entrusted with wilderness management until 1983.[7]

While each agency manages its wilderness areas independently, they all share the same values, priorities, and land use restrictions stipulated in the 1964 Act. All must strive to preserve the wilderness character of the land. According to the law this includes places "where the earth and its community of life are untrammeled by man, where man himself is a visitor who does not remain . . . retaining its primeval character and influence." This generally means a ban on resource development (e.g., mining and logging), but also restrictions on road building, motorized vehicles, mechanized transport (including mountain bikes), and permanent structures. Grazing, however, is often allowed, along with noncommercial hunting, fishing, and other recreation activities deemed compatible with management priorities.

TEXTBOX 8.1.
The Wilderness Act of 1964

Statement of Policy

Section 2 (a) In order to assure that an increasing population, accompanied by expanding settlement and growing mechanization, does not occupy and modify all areas within the United States and its possessions, leaving no lands designated for preservation and protection in their natural condition, it is hereby declared to be the policy of the Congress to secure for the American people of present and future generations the benefits of an enduring resource of wilderness. For this purpose there is hereby established a National Wilderness Preservation System to be composed of federally owned areas designated by Congress as "wilderness areas," and these shall be administered for the use and enjoyment of the American people in such manner as will leave them unimpaired for future use and enjoyment as wilderness, and so as to provide for the protection of these areas, the preservation of their wilderness character, and for the gathering and dissemination of information regarding their use and enjoyment as wilderness.

Definition of Wilderness

Section 2 (c) A wilderness, in contrast with those areas where man and his own works dominate the landscape, is hereby recognized as an area where the earth and its community of life are untrammeled by man, where man himself is a visitor who does not remain. An area of wilderness is further defined to mean in this Act an area of undeveloped Federal land retaining its primeval character and influence, without permanent improvements or human habitation, which is protected and managed so as to preserve its natural conditions and which (1) generally appears to have been affected primarily by the forces of nature, with the imprint of man's work substantially unnoticeable; (2) has outstanding opportunities for solitude or a primitive and unconfined type of recreation; (3) has at least five thousand acres of land or is of sufficient size as to make practicable its preservation and use in an unimpaired condition; and (4) may also contain ecological, geological, or other features of scientific, educational, scenic, or historical value.

Of all the different types of federal land classification, wilderness areas exude the most fully realized manifestation of environmental preservation. It most closely aligns with the passions articulated by the Romantics. Surely Henry David Thoreau and John Muir envisioned this type of nature as the ideal in their impassioned pleas for wilderness protection— landscapes devoid of the parking lots, campgrounds, and gift shops that today lie sprinkled throughout the national park system.

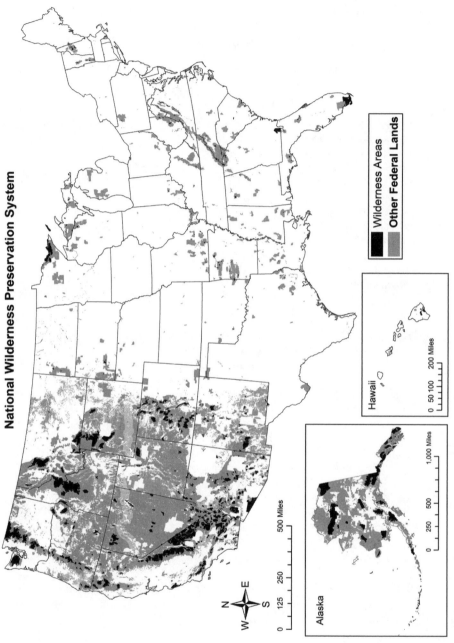

National Wilderness Preservation System

Wilderness Areas
Other Federal Lands

Hawaii

0 50 100   200 Miles

Alaska

0   250   500   1,000 Miles

N
W — E
S

0 125 250   500 Miles

Figure 8.1. The National Wilderness Preservation System. Map by Jessica Lee.

Indeed, the concept of wilderness implies a sense of pureness. Nonetheless, the "true" meaning of wilderness has been a constant source of debate, changing and evolving over time in ways that sometimes appear contradictory. While the 1964 Act gives us one definition, it did not settle disputes over its interpretation. Indeed, different understandings of what defines wilderness have been championed and deployed by various interest groups and federal agencies to advance specific agendas since the beginning.

By the same token, we can also ask if wilderness areas are, in fact, wild. If by the term *wild* we mean untamed, uncontrolled, and unmodified by human actions, do we also mean unmanaged? Is wilderness management an oxymoron? The irony is that in order to experience the freedom of "wild nature," visitors must submit to a highly regulated set of practices designed to protect and ensure the continued viability of that which is deemed wild. In short, wilderness areas are at once the most "wild" and the most regulated of all types of public lands.

This does not mean that federal managers in the course of their duties alter or modify the wilderness landscapes in ways prohibited by the 1964 Act—thinning forest stands, fighting fires, building roads—but they do attempt to control the number and behavior of human visitors. Notably, these restrictions on use appear to reflect an entirely different set of management priorities from what we've encountered so far—priorities based upon very different ways of conceptualizing nature. Nature imbued with inherent value seems paramount here. And this raises an interesting question: In wilderness, have we finally broken with the legacy of nature-as-commodity, or the perception of nature as somehow unpeopled, eternal, and pristine?

At a more pragmatic level, can we finally resolve the contradictory mandate that plagues the national parks? Are managers no longer beholden to the promotion and accommodation of recreational tourism? Do we escape the hunting and game species bias that has long overshadowed the management of national wildlife refuges (saving wildlife in order to kill it)? And can we avoid the resource development mandates and industrial influences that constantly press up against conservation measures on national forests, and which have all but crippled similar efforts on BLM lands in past decades?

Unquestionably, this is a tall order, and perhaps too much to ask of wilderness. In terms of nature-as-commodity thinking, it is clear from political debates that wilderness designation often requires passage of the "no commodity value" test. Like early advocates for the national parks, wilderness activists have had greater success when they are able to convince local congressional delegations that the land in question holds limited value for traditional resource development. Conversely, when appeals

to the inherent value of wilderness (including its worth to future generations) fall on deaf ears, proponents may revert to previously discarded
economic arguments. In so doing, they might point to environmental
services (biodiversity, watershed protection, provision of clean air, and
opportunities for public health, both physical and psychological) or stress
the local monetary gains to be had from recreational tourism.[8]

But such logic takes us right back to the contradictions folded into the
National Park Service mandate and which are echoed in the strategies
of Stephen Mather. Wilderness areas may exclude motorized recreation,
but they do in fact attract, and become magnets for, other types of nature
tourism. And in so doing, they create one of the greatest threats to their
own preservation: overuse. Already by the late 1960s, this threat was recognized by environmental organizations, which soon began publishing
books and instruction manuals on how to "leave no trace" when camping or backpacking in wilderness areas.[9] This change in thinking helped
spawn a whole new industry in gear and clothing designed specifically
for backpacking and outdoor activities.[10]

However, despite the threats posed by recreation and overuse, conceptually speaking, the greatest challenge to wilderness area protection may
have been the idea of nature as unpeopled, eternal, and pristine. Wording
contained in the 1964 Act underscores the notion of a wilderness area as a
place retaining its "primeval character," a place with only limited human
influence. The urge to protect such places by demarcating boundaries and
restricting human access appears to embrace, at least implicitly, the problematic conceptual assumption of the separation of nature and society. By
excluding humans, the task of preservation is complete. These were the
very ideas that William Cronon and others warned against back in the
1990s.[11] Taken to their logical conclusion, such thinking allows activists
to paint themselves into a corner, until there is no place left for humans
in nature. If nature can only be saved by human exclusion, what hope is
there for sustainable coexistence?

Notwithstanding these conceptual concerns, the modern history of wilderness protection suggests that most of the time, wilderness advocates
and organizations have understood the interconnections between human
and nonhuman nature. They have not proposed wilderness designation
as the *only* preferred form of public land management. As discussed below, some of the most ardent supporters of the early modern wilderness
movement argued that wilderness areas should serve as but one of many
multiple uses of the public domain, and that the land needn't be "pristine" or untouched by human hands to qualify.

Nonetheless, debates over the precise definition and meaning of wilderness, including its role in the larger environmental movement, continue
to this day. From the spiritual arguments put forth by the nineteenth-

century Romantics, to the recreational rationales offered in the 1930s and 1940s, to the ecological and economic claims espoused in more recent decades, the meaning of wilderness has constantly evolved and changed. Tracing out key moments in this history illustrates that while wilderness is not the only way humans and nature interact, it is an important way nonetheless, and it is one that provides an important focal point for working out the broader relationship between nature and American society in the public land system.

ORIGINS: THE WILDERNESS IDEA

Tracing back the idea of wilderness in American thought brings one readily to the feet of the nineteenth-century Romantics. In fact, it happens with such speed and force that it smarts just a bit.

Thoreau's thunderous dictum, "In wilderness is the salvation of the world!"[12] stands as an ear-ringing call to action. The works of Emerson, Burroughs, Catlin, Cooper, Whitman, Muir, and many others weave together artful depictions of nature with impassioned and urgent prose, providing a stalwart defense of wilderness against the intrusions of industrial society.

So it can come as a bit of a shock to learn that the modern wilderness movement was not a response to resource development on federal lands. Mining, logging, grazing . . . none of these served as the primary catalyst for action. Rather, as historian Paul Sutter explains, the threat that mobilized some of the most lucid and influential environmental minds of the day was that posed by the automobile, and in particular, leisure tourism.[13]

From the late 1910s through the 1930s, American society underwent a series of massive upheavals. But one constant amid this change was the steady growth and expansion of roads and the automobiles that used them. In 1913, five years after Henry Ford introduced the Model T, there were approximately one million passenger cars in the United States. With a human population of 97.25 million, this amounted to one car for every ninety-seven persons. By 1930, those numbers had grown to twenty-three million automobiles and just over 123 million people, providing one car for every 5.3 persons in the United States.[14] And road building kept pace, whether during the economic "boom times" of the 1920s or the New Deal back-to-work programs of the Depression Era 1930s.

In the national parks, these trends were in full evidence. The decision by Park Service director Stephen Mather to build a network of roads to and through the park system was a central part of his campaign to build a national constituency. After all, the parks needed friends to bolster congressional support, and making them was something John Muir and others had long sought. Many of Muir's writings included calls for citizens to come and visit the parks, to spend time in the mountains and recuperate from the stresses of modern life. For Mather, automobile tourism offered an effective way to achieve this goal.

However, for other park advocates, such as Robert Sterling Yard, the overwhelming success of the automobile program ushered in a new set of problems. As noted in chapter 4, by 1919, automobile visitors to national parks outnumbered train travelers four to one. While useful for transporting tourists to the parks, unfortunately, cars did not stop at the entrances. New roads allowed "auto campers" to penetrate deep into the woods, leaving scarred and littered landscapes in their wake. Meanwhile, the number of wild places for quiet solitude slowly diminished.

That same year, Yard left the NPS to serve as executive secretary of the newly formed National Parks Association,[15] a position he would hold until 1933. In his new role, Yard grew increasingly critical of the potential impacts of automobile tourism in the parks:

> Civilization has often been called the arch enemy of the primitive, but today . . . [i]t is the automobile, agent of material progress, destroyer of deserts, leveler of mountains and annihilator of time and distance that must bear that charge.[16]

The national forests exhibited similar recreational trends during this period. Threatened by the creation of the National Park Service and subsequent transfer of national forestland to expand the park system, the Forest Service aggressively promoted recreation as part of its own mandate. Constructing its first campground in 1916, the agency built roads and tourist facilities and launched a program to promote recreational visitation. Since 1905, the Forest Service had allowed private citizens to build summer cabins and lodges on aesthetically pleasing portions of national forestlands on short-term lease contracts. But with the 1915 Term Permit Act, the Forest Service now offered leases up to thirty years, attracting a new wave of visitors.

As with national parks, these developments raised concerns from national forest advocates, including those within the ranks of the Forest Service itself. Significantly, concerns focused not on the question of whether recreation should be part of the national forest mission but instead on the scale and type of recreation permitted. Like Sterling Yard, some began wondering aloud how well automobile-based recreation balanced with other forest management goals.

In 1919, the Forest Service hired Arthur Carhart as one of its first landscape architects.[17] Assigned to the regional office in Colorado, his first major project was a recreation development plan for the San Isabel

National Forest. His final plan included extensive development of cabins, resorts, trails, and roads. But it was done in such a way as to maximize public access to the most scenic sections of the forest while avoiding the monopolization of resources by those holding private leases.

Later, Carhart was assigned to Trapper's Lake, an idyllic setting in the White River National Forest. Charged with overseeing the leasing of summer cabins along the shore, Carhart once again grew concerned over the lack of public access. He concluded that the best plan was to leave the lakeshore just as it was. His final report placed cabin plots behind a half-mile buffer surrounding the lake. Carhart came to similar conclusions for a road survey project in the lake-filled Superior National Forest in Minnesota, a place that would one day become part of the Boundary Waters Wilderness on the U.S.-Canadian border. His recommendation? Leave out the roads and rely on boats as the only form of access to the area.

In 1922, Carhart left the Forest Service for a private architectural firm and devoted his time to writing about conservation matters. Though his motives were more deeply rooted in concerns for aesthetics and ensuring public access rather than wilderness preservation per se, Carhart's reports are often cited as one of the earliest calls for protecting wilderness, not only from resource development but also from the impacts of recreational tourism.

ALDO LEOPOLD AND THE FIRST WILDERNESS

By the time Carhart left the Forest Service, Aldo Leopold had already served the agency for nearly thirteen years. As chief of operations for District Three, he held responsibility for over twenty million acres of national forest in the Southwest. Born in Burlington, Iowa, schooled at Yale, and trained in Gifford Pinchot's utilitarian-oriented progressive conservation, Leopold spent his time as a forest ranger developing new ways of understanding and relating to nature (figure 8.2). Like Carhart, Leopold came to believe that portions of the national forests should be left alone entirely. And like Carhart, Leopold shared a deep belief in the democratic mission of public lands. Historians document a 1919 meeting between the two men as an encounter between kindred souls. For both Carhart and Leopold, preserving pieces of wild, untamed nature constituted a moral responsibility.

But Leopold was first and foremost a forester, not a landscape architect. In working out his early ideas about wilderness, he saw it as one among many "multiple-uses"—one that like commercial logging, mining, or grazing could serve as the "highest and best use" in some places. In one of his earliest articles on the subject, published in 1921, Leopold suggested

Figure 8.2. One of the primary architects of modern ecology and wildlife management in the United States, Aldo Leopold developed the idea of the "land ethic," and served as one of the earliest advocates for wilderness-area protection on U.S. public lands. Photo courtesy of the Aldo Leopold Foundation.

setting aside portions of the Gila National Forest from road building and commercial logging, not for its own sake, but on the basis that it constituted one of the last remaining "primitive" areas for hunting.[18]

Like many of his contemporaries, Leopold's understanding of undeveloped land was strongly informed by Frederick Jackson Turner's frontier thesis.[19] According to this view, the American ideals of independence, pragmatism, and democracy derived in no small part from the experience of struggling to survive in primitive environmental conditions. For Leopold, saving such places was profoundly patriotic insofar as they reflected a vital part of our collective national heritage.

Pristine wilderness this was not. A frontier-defined interpretation of wilderness excluded neither humans nor their impacts. Livestock grazing, hunting, subsistence logging . . . all of these were part and parcel of the frontier experience and therefore condoned. Rather, Leopold required only that large-scale commercial resource development be restricted, along with the construction of modern roads, automobiles, and the recreational activities dependent upon them.

Later in life, he would develop a philosophical basis for what he deemed to be the proper relationship between nature and human society. Expressed in concepts such as "thinking like a mountain," Leopold proposed a systems-based, biocentric worldview, in which nature and humans coexisted as interdependent partners rather than in a relationship defined in terms of "conqueror" and "conquered."[20] But in the 1920s, he focused on the pragmatic issue of finding a means to set aside some portions of national forestland from the impacts of modern society.

In 1922, Leopold crafted a formal proposal to the Forest Service Regional Office to designate portions of the Gila National Forest in New Mexico as a wilderness area. His superiors agreed, and on June 3, 1924, the Forest Service formally established the Gila as the world's first wilderness area. Though not a permanent classification, and not yet adopted as national policy, the Gila Wilderness was nonetheless a first step.

Seven years later, the Forest Service took a second step. In 1929, with strong support from Chief William Greeley and the survey work of L. F. Kneipp, the Forest Service issued the "L-20" regulation: the first federal rule allowing the formal classification of lands within a national forest system as "primitive areas." The purpose was to "maintain primitive conditions of transportation, subsistence, habitation and environment to the fullest degree compatible with their highest public use." Again, this did not preclude the use and development of timber, forage, or water resources. And it was still not a permanent classification, but dependent upon the discretion of local Forest Service personnel. Nonetheless, in this way, some fourteen million acres of national forestland received a form of wilderness protection.[21]

BOB MARSHALL AND THE WILDERNESS SOCIETY

The same year that witnessed the creation of the world's first wilderness area also marked the beginning of the remarkable Forest Service career of Bob Marshall (figure 8.3). Born to a wealthy family in New York, he spent the latter years of his youth climbing mountains. With his brother George, Marshall became the first person to scale all forty-six peaks in the Adirondack Mountains.[22] A true outdoorsman and a powerful hiker, he loved nothing more than to be immersed in wild country. He thought nothing of hiking thirty miles in a day with a heavy pack. Articulate, charismatic, with a passion for natural science and socialist causes, Marshall turned his sharp intellect to the study of forestry at the New York State College

Figure 8.3. Whether serving as a high-ranking official in the USDA Forest Service, the U.S. Department of the Interior, or as founder of the Wilderness Society, during his short life, Bob Marshall worked tirelessly to advance the cause of wilderness protection in the United States. Photo courtesy of the Wilderness Society.

of Forestry at Syracuse. After going on to complete a master's degree at Harvard in 1924, he accepted his first Forest Service appointment at the Northern Rockies Research Station in Missoula, Montana.

During his time in Montana, Marshall developed a passion for wilderness preservation. In 1930, while completing his PhD in plant physiology at Johns Hopkins University, Marshall published an article in *Scientific Monthly* titled "The Problem of the Wilderness." In it, he made a strong case for wilderness protection, offering a definition that emphasized not only aesthetic values, but the opportunities wilderness provided for self-reliance:

> I shall use the word *wilderness* to denote a region which contains no permanent inhabitants, possesses no possibility of conveyance by any mechanical means and is sufficiently spacious that a person in crossing it must have the

experience of sleeping out. The dominant attributes of such an area are: first, that it requires anyone who exists in it to depend exclusively on his own effort for survival; and second, that it preserves as nearly as possible the primitive environment. This means that all roads, power transportation and settlements are barred. But trails, temporary shelters, which were common long before the advent of the white race, are entirely permissible.[23]

Marshall's vision shared much with Leopold, reflecting the influence of Turner's thesis. But his argument for wilderness did not end with the need to preserve a frontier experience. He went on to describe the health benefits of hiking outdoors, including the mental or psychological gains from relaxation and testing oneself. He quoted Bertrand Russell, claiming that if men could "risk their lives" scaling mountains, they might lose the yearning for war. Marshall defended wilderness as a "minority right" against the tyranny of the "automobile majority" and as a healthy component of democratic society. The economic losses from logging and road building would be more than offset by the economic gains of wilderness recreation, available to all free of charge. He echoed Leopold's early plea for a national survey of lands suitable for wilderness protection and concluded with a call to action that foreshadowed the creation of the Wilderness Society:

There is just one hope of repulsing the tyrannical ambition of civilization to conquer every niche on the whole earth. That hope is the organization of spirited people who will fight for the freedom of the wilderness.[24]

After spending a year in Alaska, Marshall returned home to write the best-selling *Arctic Village*. In this ethnographic study of Wiseman, a small town in northern Alaska, Marshall extolled the virtues of life in a community closely intertwined with the wilderness. Then, in 1933, Marshall transferred to the Interior Department to become head of the Division of Forestry and Grazing in the Office of Indian Affairs. In this role, Marshall pushed for wilderness declarations on Indian reservations, arguing that such areas might better protect native cultures from the intrusions of modern society. He also lobbied his new boss, Interior Secretary Harold Ickes, to expand wilderness protections in the national parks. Meanwhile, Marshall continued his call for a nationwide wilderness inventory led by a planning board with backing from Congress.[25]

That same year, Marshall published *The People's Forests*, in which he made the case for the federal control of all forestlands in the United States. Convinced that private ownership was leading to the degradation of our remaining timberlands as companies sought to maximize short-term profits, Marshall argued that only scientific management carried out by federal managers could possibly stem the tide of waste and exploitation.

Despite public sympathies for such a position in the midst of the Depression, the book unsurprisingly drew harsh criticism, not only from the logging industry but from the Forest Service as well.[26]

Though his book did little to sway national forest policy, some individuals within the agency shared Marshall's socialist leanings. One of these was Benton MacKaye, a forester turned regional planner. In his work, MacKaye sought opportunities for blending wilderness protection with sustainable resource development in order to assist lower-income Americans wishing to settle on the public domain. In his 1921 article titled "An Appalachian Trail: A Project in Regional Planning," MacKaye called for creation of a footpath stretching from New Hampshire to North Carolina (later expanded to a two-thousand-mile trail from Maine to Georgia).[27]

His original intention for the path was to connect a series of progressive camps or settlements. These were to be designed not only for recreation but also for the establishment of working cooperative communities dependent upon the sustainable development of resources in the Appalachian region. Over time, as local support for the recreational aspects of the trail grew, the goal of establishing cooperative settlements began to fade. Before long, the Appalachian Trail became viewed first and foremost as a recreational pathway, connecting wild lands along the crest of the mountain range. Even MacKaye began referring to it as a "wilderness way" through civilization.[28]

By 1934, the Appalachian Trail was almost complete, but threatened in sections by the intrusion of new highways. In fact, it was concern over one such highway, the proposed Blue Ridge Parkway, that finally brought together the founders of the Wilderness Society. Designed as a "skyline highway" linking Shenandoah and the Great Smoky Mountains National Parks, MacKaye worried that the new road would not only negatively affect the national parks but also large sections of the Appalachian Trail.

In October, MacKaye and his colleagues, Bernard Frank and Harvey Broome, attended the American Forestry Association conference in Knoxville, Tennessee, where Bob Marshall was scheduled to speak. During the meeting, they met up with Marshall and decided to travel together on a conference-sponsored trip to inspect a nearby Civilian Conservation Corps (CCC) work camp. En route, discussion turned to the formation of an advocacy group for wilderness protection based on Marshall's 1930 proposal. Conversation grew so animated that they pulled off to the roadside for a more focused discussion. Finally, they reached agreement on a core set of principles for a new organization that would become the Wilderness Society.[29]

Marshall then sent invitations to a number of other leading wilderness advocates to join the group. In 1935, the eight founding members included Bob Marshall, Aldo Leopold, Robert Sterling Yard, Benton

MacKaye, Ernest Oberholtzer, Harvey Broome, Bernard Frank, and Harold Anderson. Oberholtzer had been recently appointed by President Franklin D. Roosevelt to chair the Quetico-Superior Lake Committee, a government task force whose work led to the protection of the Boundary Waters Canoe Area. Broome was president of the Great Smoky Mountain National Park Conservation Association. Frank, like MacKaye, was a Tennessee Valley Authority forester, and Anderson was a prominent member of the Potomac Appalachian Trail Club.

They kept membership intentionally small in order to guard against the erosion of their ideals. The purpose of the Wilderness Society was nothing less than permanent federal protection for large unroaded areas of undeveloped land as wilderness. As the energetic and informal leader of the group, Marshall used his personal fortune to bankroll the organization. However, due to his status as a federal employee, Marshall declined the presidency. Marshall looked to Leopold as the true leader of their cause, but he had just taken a faculty position at the University of Wisconsin.[30] Leadership thus went to Robert Sterling Yard, who served as Wilderness Society president until his death in 1945.

In the first years following its formation, Society members pushed for wilderness protections largely from within their individual spheres of influence. Meanwhile, Yard began publishing the Society's magazine, *The Living Wilderness*, making the case for wilderness protection to a broader public audience. In 1937, Marshall returned to the Forest Service to lead the Division of Recreation and Lands. In his new post, Marshall successfully garnered L-20 protections for some fourteen million acres of national forestlands. He also personally financed a survey of remaining roadless areas larger than three hundred thousand acres within the national forest system. The survey identified nearly forty-five million acres of lands suitable for potential wilderness classification.

Meanwhile, under the urging of Interior Secretary Ickes, the National Park Service also advanced its thinking about wilderness protection. For example, in 1938, Olympic National Park in Washington State was established as a "wilderness park" carved out of Forest Service L-20 primitive areas. That same year, Ickes proposed another such park in Kings Canyon in California. Congressional debates were heated, as was the interagency rivalry between the Forest Service and Park Service over who could best manage wilderness areas.

Although he was once again a Forest Service official, during congressional hearings Marshall and the Wilderness Society endorsed the National Park Service over the Forest Service as the preferred manager of wilderness lands.[31] While this view essentially meant national park expansion at the expense of the national forest system, Marshall, nonetheless, continued to work tirelessly to strengthen Forest Service designations to ensure better protection of the wilderness that remained within the national forest system.

To this end, in 1939, Marshall helped to develop a new set of Forest Service rules for wilderness management. The new "U Regulations" identified a three-tiered classification scheme for wilderness protection in the national forest system.[32] U-1 tracts consisted of areas larger than one hundred thousand acres where, for the first time, commercial logging was explicitly prohibited. They were established by the Secretary of Agriculture upon recommendations from the chief forester. Significantly, only the Secretary could modify or delist such areas, and any changes required a process of public hearings and notifications. These safeguards made it much less likely that wilderness designations might be reversed over time (as was the case with the old L-20 rules). The U-2 category was for smaller areas of five thousand to one hundred thousand acres. Land use restrictions remained the same, but the chief forester held final say in their designation and modification. Finally, the U-3 classification applied to any size tract. Once again, roads were prohibited; however, commercial logging and other resource development was allowed.

Despite the gains made in adopting the U-Regulations, their implementation by the Forest Service was painfully slow. After twenty-five years, over one-third of L-20 wilderness lands (some 5.5 million acres) had still not undergone reclassification.[33] Moreover, during that same time period, designated wilderness acreage grew only by 2.8 percent, from 14.2 million to 14.6 million acres. The system was essentially stagnant. But even more worrisome were boundary changes that severely impacted the composition of lands receiving protection. Each time L-20 lands were reassessed under the new U-Regulations, foresters took the opportunity to adjust boundaries to remove timber-rich lands for logging. These acres were then replaced with nonforested and less biologically diverse lands, including rocky slopes and high-elevation mountain peaks.[34]

Shortly after the adoption of the U-Regulations in 1939, Bob Marshall unexpectedly died of a heart attack at the age of thirty-eight. He bequeathed his fortune to three causes: socialism, civil liberties, and the preservation of wilderness. Approximately $400,000 went to establish the Robert Marshall Wilderness Trust, a crucial asset that ensured the fiscal viability of the Wilderness Society for years to come.[35] In 1940 the Forest Service designated the Bob Marshall Wilderness in Montana in recognition of his contributions to wilderness protection. That same year, Congress established Kings Canyon National Park by renaming and expanding General Grant National Park in California with wilderness areas taken from national forestlands.

HOWARD ZAHNISER AND THE WILDERNESS ACT OF 1964

In 1945, Robert Sterling Yard passed away, and Benton MacKaye became the new president of the Wilderness Society. The position of executive secretary fell to Howard Zahniser, who also assumed the editorship of *The Living Wilderness* (figure 8.4). The central concern during the 1940s was the temporary status of wilderness areas within the federal land system. Both the national parks and national forests retained administrative discretion over the future status of such places. Hence, a change in leadership could signal the end of U-Regulation, L-20, or "wilderness park" status at any time. The lack of a common definition or set of established management goals only added to the uncertainty.

Figure 8.4. Howard Zahniser served as executive secretary of the Wilderness Society and was the principal author of the 1964 Wilderness Act. Though he did not live to see the bill become law, Zahniser's impassioned and determined advocacy for the bill on Capitol Hill and elsewhere played a significant role in its eventual success. Photo courtesy of the Wilderness Society.

It is telling that for the next twenty-five years after Bob Marshall's death, no new wilderness areas were established in the national forest system. Despite calls by Leopold (1925), Marshall (1934), and Secretary Ickes (1939) for a systemic review of lands for wilderness designation, the status quo remained essentially unchanged. Much of the work of the Wilderness Society, Sierra Club, and other environmental groups during this period was defensive in nature, trying to protect existing wilderness from the threat of development or reclassification rather than seeking to expand to new areas.

In 1947, the Wilderness Society decided to adopt a new "offensive strategy" by pursuing congressional legislation for wilderness protection. In support of this goal, the Sierra Club began hosting a biennial wilderness conference in 1949. However, before this campaign could gather much steam, a new threat appeared on the horizon. In 1950, the Bureau of Reclamation announced plans to construct a series of dams on the Colorado Plateau. One of these was slated for Echo Park, a remote place along the Green River in northeastern Colorado that also happened to be located within the boundaries of Dinosaur National Monument.

For environmental activists it was Hetch Hetchy all over again. Without the aid of modern environmental laws, the only viable strategy to stop the dam was to sway public opinion and somehow convince Washington to reject the plan, just as John Muir had attempted to do half a century earlier. They formed a coalition that included the National Parks Association; the Wilderness Society, now led by Olaus Murie; and the Sierra Club under the leadership of its first executive director, a young firebrand named David Brower. Working together for the first time, the coalition partners lobbied Congress and mounted public relation campaigns, both in the national media and at the grassroots level.

Once again, Bernard DeVoto lent his considerable influence to the cause, penning an essay titled "Shall We Let Them Ruin Our National Parks?"[36] Reaching a national audience, DeVoto's writing proved to be a powerful tool in shaping public opinion. In 1953, Brower floated down the river with writer Wallace Stegner, who afterward penned *This Is Dinosaur: The Echo Park Country and Its Magic Rivers*. Published by Alfred Knopf, the book's eight essays and magnificent photographs were distributed to every member of Congress as well as to newspapers and conservation groups across the country, echoing the *National Parks Portfolio* campaign devised by Stephen Mather decades before.

Brower and other campaign leaders strategically cast their arguments in a systems-based framework. Any infringement on the integrity of Dinosaur National Monument, the logic went, damaged the integrity of the entire national park system. They also pointed out that enlarging dams at Glen Canyon and Flaming Gorge would offset water storage losses from

Echo Park. After a five-year effort, in 1955, the coalition succeeded in convincing federal officials to stop the dam project and adopt a provision protecting national park system lands from any future dams.

For Brower, the victory was bittersweet given the environmental harm that resulted from the expansion of the Glen Canyon and Flaming Gorge dams. He vowed to never again allow the sacrifice of one area in order to save another. Nonetheless, the Echo Park fight taught activists new skills in managing a national campaign. By working together as a coalition to achieve a shared goal, environmental groups learned how to effectively lobby members of Congress and mobilize public support. With this new-found confidence, Zahniser went to work drafting a wilderness bill.

In 1956, after seventeen drafts, the bill was finally ready for Congress. Senator Hubert Humphrey, a Democrat from Minnesota, agreed to sponsor the bill in the Senate, while Representative John P. Saylor, a Republican from Pennsylvania, brought it forward in the House. Despite bipartisan support, the bill triggered immediate opposition from the Eisenhower administration, including criticism from the leadership of the Forest Service and the National Park Service. Both agencies saw the bill as an affront to their expertise and authority to manage wilderness as they saw fit. Western timber, grazing, and mining interests likewise protested, throwing the full might of their political influence against the bill. Consequently, early versions of the bill suffered a quick death.

Not to be deterred, over the next eight years Zahniser continuously drafted new versions of the bill. All the while, environmental groups persistently lobbied members of Congress for their sponsorship and support. During the 1960 presidential campaign, Senator John F. Kennedy vowed to support a wilderness bill if elected. Once in office, the new administration made good on that promise. Interior Secretary Stewart Udall and Agriculture Secretary Oliver Freeman both endorsed the wilderness bill. Environmental activists stepped up their lobbying efforts, sending every member of Congress a copy of Wallace Stegner's eloquent "wilderness letter" that he had submitted in earlier testimony (textbox 8.2).

Getting the bill through the House, however, required a different approach. In particular, it demanded compromise with the powerful House Interior and Insular Affairs Committee chairman, Representative Wayne Aspinall. A Democrat from Colorado, Aspinall had long supported western resource development interests and was adamantly against the idea of a new wilderness law. Zahniser's original bill called for creation of a "wilderness committee" that would make recommendations to the president on wilderness designations. With presidential approval, the recommended lands would then automatically gain wilderness area status. But Aspinall wanted to ensure that Congress had a greater say in the process. He demanded, therefore, that an act of Congress be required for each

TEXTBOX 8.2.
Wallace Stegner: The Wilderness Letter

Something will have gone out of us as a people if we ever let the remaining wilderness be destroyed; if we permit the last virgin forests to be turned into comic books and plastic cigarette cases; if we drive the few remaining members of the wild species into zoos or to extinction; if we pollute the last clear air and dirty the last clean streams and push our paved roads through the last of the silence, so that never again will Americans be free in their own country from the noise, the exhausts, the stinks of human and automotive waste. And so that never again can we have the chance to see ourselves single, separate, vertical and individual in the world, part of the environment of trees and rocks and soil, brother to the other animals, part of the natural world and competent to belong in it. Without any remaining wilderness we are committed wholly, without chance for even momentary reflection and rest, to a headlong drive into our technological termite-life, the Brave New World of a completely man-controlled environment. We need wilderness preserved—as much of it as is still left, and as many kinds—because it was the challenge against which our character as a people was formed. The reminder and the reassurance that it is still there is good for our spiritual health even if we never once in ten years set foot in it. It is good for us when we are young, because of the incomparable sanity it can bring briefly, as vacation and rest, into our insane lives. It is important to us when we are old simply because it is there—important, that is, simply as an idea.

Source: Wallace Stegner, selection from "Coda: Wilderness Letter," in *The Sound of Mountain Water* (New York: Doubleday, 1969), 146–7. The full letter was submitted as testimony to the U.S. Outdoor Recreation and Resources Review Commission on December 3, 1960.

new wilderness area. In addition, he insisted that new mining claims be allowed in Forest Service wilderness areas for the next twenty years (until December 31, 1983).

The bill's sponsors relented. And in 1964, with Aspinall's terms now included in the bill, the Wilderness Act finally passed through Congress. On September 3, 1964, President Lyndon B. Johnson signed it into law at a ceremony in the Rose Garden (figure 8.5). Sadly, neither Zahniser nor Murie were able to attend. Olaus Murie passed away in 1963, and Howard Zahniser died in May 1964 at the age of fifty-eight. Only one week after his final congressional testimony on the bill, Zahniser finally lost a long battle with a heart condition. Nonetheless, forty years after the creation of the first wilderness area in the Gila National Forest, the United States had its first statutory wilderness protection law.

Figure 8.5. The signing ceremony for the 1964 Wilderness Act took place in the White House Rose Garden on September 3, 1964. President Lyndon B. Johnson was joined by Alice Zahniser and Margaret Murie, the spouses of Howard Zahniser and Olaus Murie, respectively. Photo credit: Abbie Rowe, courtesy of the National Park Service Historic Photograph Collection, Harpers Ferry Center.

THE WILDERNESS ACT IN PRACTICE

The Wilderness Act designated the first 9.14 million acres of wilderness areas in the United States under federal law. Most of the fifty-four newly established units were drawn from national forest U-Regulation lands. They included the Boundary Waters Canoe Area in Minnesota, the Bob Marshall Wilderness in Montana, the John Muir Wilderness in California, and fittingly, the Gila Wilderness in New Mexico (table 8.1).

In addition to designating new wilderness areas, the Act also: 1) established the National Wilderness Preservation System, 2) created a process for the designation of new wilderness, and 3) offered the first legal definition of the term. But as we will see, it did not settle debates over the final meaning and interpretation of wilderness. Additionally, the Wilderness Act mandated federal land agencies to systematically inventory their lands for future wilderness area designations. These inventories were to be completed within ten years and include agency reviews and public

Table 8.1.   Milestones: The First Wilderness Areas

Date	Type	Name (State)
1924	First Wilderness prior to 1964 Act	Gila (NM)
1964	Designated by 1964 Wilderness Act[a]	Fifty-four different areas covering over 9.1 million acres, including Boundary Waters Canoe (MN), Gila (NM), Bob Marshall (MT), and John Muir (CA)
1968	First National Wildlife Refuge Wilderness Area	Great Swamp (NJ)
1968	First National Forest Wilderness after 1964 Act	San Rafael (CA)
1970	First National Park Wilderness after 1964 Act	Craters of the Moon (ID)
1972	First de facto Wilderness Area[b]	Scapegoat (MT)
1983	First BLM Land Wilderness	Bear Trap Canyon Unit of Lee Metcalf (MT)

[a]These wilderness areas were derived from national forest lands. According to the Wilderness Act, they were previously designated by the Forest Service as either "wilderness," "wild," or "canoe" areas, and they had held that status in the thirty days prior to the law's enactment.
[b]De facto refers to the fact that this was not one of the areas required by the Wilderness Act for review by federal agencies.

hearings, ultimately leading to presidential recommendations to Congress for legislative action.

The National Park Service and U.S. Fish and Wildlife Service received specific orders to evaluate all remaining roadless areas within their respective jurisdictions. In contrast, the Act required the Forest Service to assess only the roughly five million acres of L-20 lands that had not yet received reclassification under the U-Regulations. Significantly, the act was silent regarding the remaining sixty million acres of roadless areas in the national forest system, and it contained no mention whatsoever of the lands administered by the Bureau of Land Management.

For many wilderness advocates, however, the greatest concern was the compromise made with Congressman Aspinall. The requirement of an act of Congress for each new wilderness area seemed like an impossible obstacle that fundamentally weakened the final law. However, in time, this provision proved to be one of the Wilderness Act's greatest strengths.[37] Although congressional statutes can be difficult to pass, they are also much more difficult to overturn than the discretionary rules of federal agencies. Hence, once declared it was highly unlikely that the designation of a new wilderness area would be reversed. But even more importantly, the legislative process opened the door for citizen involvement. Citizen groups could now propose wilderness area designations directly to Congress without waiting on federal agencies. This was terribly important since most of the federal land agencies were in no hurry to reclassify their lands as wilderness.

According to Doug Scott, a leader in the Wilderness Society's national campaigns during this period, only the U.S. Fish and Wildlife Service undertook wilderness inventories with any sort of enthusiasm.[38] In contrast, both the Park Service and Forest Service moved slowly. Viewing the entire enterprise as a process leading to the direct loss of discretionary power, both agencies looked for opportunities to bend the task to their advantage.

A tactic commonly deployed by the National Park Service involved a "Swiss cheese" approach to wilderness designation.[39] By demarcating "buffer zones" around potential wilderness areas and identifying areas of exclusion within them, Park Service officials argued they could better protect wilderness areas while providing places where "near wilderness" experiences might be enjoyed by park visitors less equipped or interested in rugged backcountry hiking. For critics, these buffers and exclusion zones amounted to nothing more than recreational "holes" in the wilderness "cheese." They effectively shrank and fragmented wilderness areas while providing park officials the option of carrying out future development projects to promote recreational tourism.[40]

Such tactics were exemplified in Isle Royale National Park, one of the first national parks reviewed for wilderness designation. For this island located near the north shore of Lake Superior, the Park Service ultimately recommended that 119,612 acres of the total 134,000 acres be designated as wilderness. While this appeared to be an exceptionally generous proposal, critics pointed out that the exclusions were not necessary and set a dangerous precedent.[41]

The Forest Service took a different approach to the matter. Adopting a "purity" definition of wilderness, agency leadership argued that forest lands exhibiting any significant human modification in the past were ineligible for wilderness designation under the 1964 Wilderness Act.[42] This position was first expressed in 1968, when the first new national forest wilderness area was established. Carved out of an old L-20 primitive area on the Los Padres National Forest located near the coastal city of Santa Barbara, California, the San Rafael Wilderness had been protected since 1932 because it provided habitat to the threatened California condor. After an internal review and the mandated public hearings, the Forest Service recommended expanding the original seventy-four-thousand-acre primitive area into a 142,700-acre wilderness area in order to cover a greater portion of the condor's habitat. A point of contention arose, however, when conservationists argued for the inclusion of a one-thousand-acre zone in which Forest Service roads had been constructed as fire breaks to stop a recent fire. Although the zone fell within the condor's habitat, the Forest Service would not relent due to the fact that it contained roads.

In 1971, this "purity" approach to wilderness took on even greater meaning when the associate chief forester, John McGuire, declared at the Sierra Club biennial wilderness conference that no "pure" wilderness areas remained in the eastern United States. Instead, he offered a new Forest Service–defined category of "wild areas" to encompass lands that had been modified historically but could still be candidates for special preservation-based management. However, for environmental activists, this dual approach to wilderness designation posed a significant threat to the integrity of the system. They soon responded with a number of political strategies to expand wilderness in the eastern United States (discussed below).

In contrast to both the National Park Service and the Forest Service, the U.S. Fish and Wildlife Service adopted what came to be known as a "creative approach" to wilderness designation. Environmentalists welcomed this much more flexible response to the Wilderness Act, as illustrated in the first wilderness area established in the national wildlife refuge system: at the Great Swamp National Wildlife Refuge in New Jersey. As explained by historian James Turner, the refuge was originally established through local activism. As part of a grassroots campaign led by homemaker Helen Fenske to stop construction of a new airport, local residents successfully purchased the land. The group then donated the property to the Interior Department to create the Great Swamp National Wildlife Refuge.[43]

Located near major highways and residential developments, the land was littered with abandoned buildings and bifurcated by a power line and an old, crumbling paved road. In short, the refuge appeared to have little in common with the definition of wilderness laid out in the 1964 Act. The law called for protecting roadless areas of at least five thousand acres or of manageable size and to provide opportunities for solitude and "primitive" wilderness experiences. But the Great Swamp appeared to offer none of these things. Not to be deterred, U.S. Fish and Wildlife Service managers pushed ahead and recommended a 2,400-acre tract on one side of the old road. Though much smaller than the five thousand acres called for in the Act, the area was nevertheless deemed "manageable" as wilderness. When local activists pushed for inclusion of the area on the other side of the road (including the road itself), the agency agreed to carry the proposal forward.

In congressional debates, wilderness opponents worried that Great Swamp might set a dangerous precedent, stretching the definition of wilderness in ways that might prevent contentious development projects in the future. Nonetheless, pro-wilderness supporters carried the day by arguing that the land could not only be restored, but that in so doing, it would protect local watersheds and provide a much needed respite from the congested, urbanized landscapes of New York City and northern New Jersey. In the end, the road was revegetated, the buildings removed, and the Great Swamp Wilderness became the first such designation in the national wildlife refuge system.

In addition to the mandated self-studies carried out by federal agencies, new wilderness proposals also began to emerge from citizen groups who petitioned Congress directly. Environmentalists soon found that to be most effective in these campaigns, a reorganization of the wilderness movement was in order. Rather than focus attention and resources on lobbying in Washington, D.C., as they had done for the 1964 Act, environmental groups began directing their energies at the local and regional level. Members of local groups and chapters were best situated to know the qualities of local federal lands and thereby draft the most effective citizen proposals. Mobilizing at the grassroots level, local activists presented their plans to their respective congressional delegations for sponsorship. Rather than lead the fight, national organizations now played a supporting role in these increasingly local initiatives.

In 1968, one of the first efforts to plan a citizen proposal for a wilderness area on America's public lands began in Montana. After a four-year process, in 1972, the Scapegoat Wilderness became the first de facto roadless area to receive wilderness area designation in the United States. Carved out of national forestlands, this achievement was not only applauded by environmentalists but also captured the full attention of the Forest Service.

WILDERNESS AND THE NATIONAL FORESTS

While pleased with these early successes, wilderness advocates remained concerned about the inventory process and, especially, the sixty million acres of national forest roadless areas still subject to Forest Service discretion. Absent a mandate to evaluate these lands, they remained vulnerable to resource development. For its part, the Forest Service was loath to surrender its discretionary authority over any lands within the national forest system. With the adoption of its "purity" approach to wilderness, few places on the national forest appeared to qualify for designation anyway.

Nonetheless, the growing success of environmental groups in turning the legislative process to their advantage on the question of wilderness designation spurred the agency into preemptive action. Though only required to inventory some five million acres of forestland under the

Wilderness Act, in 1971, Forest Service chief Edward Cliff initiated a survey of fifty-six million acres of remaining roadless areas for potential wilderness designation, titled the Roadless Area Review and Evaluation (RARE). After two years of study and a series of hastily arranged public hearings, the agency issued its final report in January 1973. Of the total number of acres studied, the Forest Service recommended 12.3 million acres for wilderness designation.[44]

Immediately, the Sierra Club filed suit, arguing that the study was rushed, superficial, and lacked proper adherence to NEPA requirements calling for environmental impact statements (EISs). The court issued an injunction halting all logging in national forest roadless areas until hearings could take place in November 1972. Rather than fight the legal allegations, the Forest Service decided to settle with the Sierra Club out of court. In exchange for dropping the lawsuit, the Forest Service agreed to comply with NEPA, completing an environmental assessment before conducting any resource development in roadless areas. It also agreed to stop all timber sales slated for roadless areas in 1973.

Meanwhile, with backing from environmental groups, Congress passed the Eastern Wilderness Areas Act in 1975.[45] Directly addressing the Forest Service's "pure" definition of wilderness, the law explicitly tossed out this interpretation for wilderness inventories in national forests and national parks. Instead, it directed the agencies to assess their lands with a much broader interpretation of wilderness, one that included past modifications that might be mitigated or restored over time. Additionally, the law created sixteen new wilderness areas in national forestlands east of the Mississippi River covering over two thousand acres, and it identified seventeen areas for further wilderness study.

A second congressional critique of Forest Service resistance to wilderness protection took the form of the 1978 Endangered American Wilderness Act. The Act created thirteen new wilderness areas and enlarged four existing areas, protecting a total of 1.3 million acres. Wording in the law contained a direct rebuke of Forest Service policy, declaring that these lands were "not being adequately protected or fully studied for wilderness suitability by the agency responsible for their administration."[46]

Congressional hearings for the Endangered American Wilderness Act had begun in 1976 and provided a forum for airing a series of stinging critiques against the Forest Service. In response to these hearings, and with strong support from the new Carter administration, the agency embarked on a second wilderness inventory, christened RARE II. For this study, the Forest Service dropped the "purity" definition and expanded its assessment to over sixty-two million acres of roadless areas. Forest managers also attempted to be more objective, adopting new techniques that sought to quantify wilderness and other multiple-use characteristics through a variety of metrics, including cost-benefit analysis. However, for wilderness activists, the final recommendation of some fifteen million acres for wilderness designation differed only slightly from the original RARE I study.[47] Once again, environmentalists charged that the assessment process was faulty. Because the assumptions built into the quantitative models misrepresented actual conditions on the ground, the findings were inaccurate.

Just as with RARE I, the Forest Service's RARE II recommendations were met with a lawsuit, this time brought by the state of California. The federal court ruling, which was upheld on appeal, ordered the Forest Service to conduct area-specific environmental impact statements before any roadless area in a national forest could be developed or used for non-wilderness purposes. Though the ruling only applied to national forests in California, timber and mining interests feared it might soon apply to other states. For this reason, they agreed to work with environmental groups to craft state-by-state legislation that would "release" portions of national forest roadless areas for resource development in exchange for new wilderness designations in other parts of the forest. This move opened the door to a steady stream of federal laws expanding and establishing new wilderness areas on a state-by-state basis for the next decade.

In negotiations over these bills, wilderness advocates pushed for "soft release" language that allowed areas to be reconsidered for wilderness designation in the next round of forest planning, as opposed to "hard releases" that permanently removed certain lands from wilderness consideration in the future.[48] The challenge was orchestrating these compromises not only with congressional opponents but also between environmental leaders in Washington, D.C., and the local grassroots members of environmental groups in individual states. In some cases, the national compromises with the timber industry, politicians, and Forest Service called for less wilderness than local activists wanted (as was the case in California and Oregon), and sometimes more (as was the case in New Hampshire). As a result, the 1980s witnessed a flurry of statewide wilderness bills, all of them signed into law by the most unlikely of U.S. presidents: Ronald Reagan. During this decade, Congress passed over twenty state-based federal wilderness laws, enlarging the system by approximately eight million acres.

EXPANDING THE WILDERNESS SYSTEM

Since the 1970s, despite growing partisanship in Congress, the National Wilderness Preservation System has continued to expand by leaps and bounds. In 1976, the Federal Land Policy and Management Act (FLPMA) ordered the Bureau of Land Management to finally join the other federal land management agencies in assessing its holdings to identify wilderness study areas. And in 1980, the single largest statutory expansion of the wilderness system in history occurred with passage of the Alaska National Interest Lands and Conservation Act (ANILCA). After an arduous decade-long political battle (chapter 3), in a single stroke, ANILCA designated 56.4 million acres of new wilderness areas in a variety of public land types in Alaska, nearly tripling the size of the system. That same year marked the creation of the largest single wilderness area in the lower forty-eight states: the 2.3 million-acre Frank Church-River of No Return Wilderness in Idaho.

Following on the heels of the wilderness expansions of the 1980s, Congress kicked off the new decade with passage of the Tongass Timber Reform Act of 1990. This law set aside three hundred thousand acres of remaining old-growth stands as wilderness in Alaska's Tongass National Forest. And in 1994, the California Desert Protection Act added 7.7 million more acres to the national wilderness preservation system. At the end of the decade, the Forest Service's lackluster efforts with RARE I and II were finally addressed with a completely new roadless area inventory and assessment, leading to the Roadless Area Conservation Rule. After a decade of legal battles, the rule succeeded in protecting over fifty-eight million acres of national forest lands from road construction and large-scale commercial logging. While falling short of full statutory wilderness classification, these lands nonetheless enjoy similar regulatory protection under the auspices of Forest Service administrative rules (chapter 5).

With the dawn of a new century (and millennium), Congress has continued to pass a fairly steady series of laws designating new wilderness areas, despite sharp ideological differences between the two most recent presidential administrations. These laws have established new wilderness areas and expanded existing ones, in states ranging from California (318,000 acres) to New Hampshire and Vermont (76,000 acres), and from Nevada (over one million acres) and Utah (100,000 acres) to Wisconsin (35,000 acres). Even U.S. territories, such as Puerto Rico, have received new wilderness area designations (10,000 acres). In 2009, President Obama signed the Omnibus Public Land Management Act that established an additional two million acres of wilderness in the system, including 700,000 acres in California, 250,000 acres each in Utah and Colorado, and approximately 80,000 acres in Appalachia. As of 2013, there were some twenty bills pending in Congress to establish another two million acres across eleven states.

While partisan politics still define—and hinder—long-running efforts to establish wilderness designations in certain places (in ANWR, for example, or Yellowstone National Park), the steady list of successes elsewhere seems to affirm the benefits of a statutory approach to wilderness designation. At first glance, this formal top-down regulatory approach also appears to explicitly reject collaborative management as a preferred decision-making or conflict resolution tool in contrast to managerial trends on other types of public lands. However, the collaborative moment in wilderness protection is still present. It can be found quite readily at the grassroots level, as citizens and activists work to build local coalitions of support for state-based proposals, many of which include wilderness creation or expansion in exchange for opening other public lands for resource development. It also occurs after designation, during comprehensive planning processes for wilderness areas lying within specific national wildlife refuges, national forests, or national parks. But there are other forms of wilderness that by their very nature defy the boundaries of public lands. In so doing, they present unique conceptual, political, and physical challenges to both federal land agencies and environmental organizations—challenges that demand collaborative responses if they are to continue to play an essential role in America's public land system.

CASES

The very same sentiments that launched the wilderness movement in the United States played a significant role in other campaigns to protect and preserve America's public lands. Efforts to establish scenic hiking trials and protect wild and scenic rivers both illustrate this point. While the roots of the trails movement arguably predate the wilderness movement (or at least the 1935 formation of the Wilderness Society), proponents have long characterized these footpaths as a means of connecting and experiencing American wilderness in a sustained and meaningful way. Similarly, the idea of wild and free-flowing rivers denotes a sense of untamed nature that we often associate with wilderness areas.

But at the same time, the physical form and extent of these trails and waterways draw attention to, and generate appreciation for, the systemic character of ecosystems, including the fundamental ecological services they provide to human society. Like wilderness, these characteristics challenge and complicate the dominant notion of nature-as-commodity, as well as conceptions of nature as unpeopled, eternal, and pristine. Moreover, the degree to which these trails and rivers stretch beyond and between "islands" of federally owned land puts them at greater risk of resource development. They are placed on the front lines of debates not only about wilderness but also about the physical and conceptual meaning of public and private space.

Wild and Scenic Rivers

The idea of wild rivers has a long history in the environmental movement. The earliest and most famous expressions derive from campaigns against dam projects. Battles over Hetch Hetchy, Echo Park, and the Grand Canyon denote historic moments that helped define the modern environmental movement. In each instance, activists focused on the landscape impacts of large-scale flooding to create reservoirs. However, for the general public—aside from the devastation caused by dam failures—the idea that dams might be problematic for the environment was much slower in coming. For decades, large-scale dam construction was considered a marvel of American ingenuity, something that benefitted society in broad strokes by providing essential water resources, clean electricity, and flood protection. In the 1930s, completion of massive dam projects such as Hoover on the Colorado River and Bonneville and Grand Coulee on the Columbia River, inspired hope for a better future for many Americans. Unless one lived nearby or depended on these free-flowing rivers for purposes of livelihood or community well-being (commercial fishermen or indigenous peoples reliant upon salmon runs, for instance), the negative effects were not often understood.

It was not until the early 1960s that the ecological science caught up with the general public, offering explanations and evidence of the detrimental effects of dams, including loss of habitat, negative impacts upon migrating fish, changes in water temperature, salinization, and sedimentation buildup. In 1962, the Outdoor Recreation Resources Review Commission (ORRRC) recommended that some free-flowing rivers and streams in the United States be preserved on the rationale that their "natural scenic, scientific, aesthetic, and recreational values outweigh their value for water development and control purposes now and in the future."[49] To identify such rivers, the ORRRC formed the Wild Rivers Committee to study the issue and make recommendations to Congress. Sponsored by Senator Frank Church (D-Idaho) and Representative John P. Saylor (R-Pennsylvania), the Wild and Scenic Rivers Act was signed into law by President Lyndon B. Johnson on October 2, 1968. The Act identified eight rivers for protection,[50] and stated the following:

> It is hereby declared to be the policy of the United States that certain selected rivers of the Nation, which, within their immediate environments possess outstandingly remarkable scenic, recreational, geological, fish and wildlife, historic, cultural or other similar values, shall be preserved in a free-flowing condition, and that they and their immediate environments shall be protected for the benefit and enjoyment of present and future generations.[51]

To achieve these ends, the law prohibited dams and called for a quarter-mile protected buffer zone on either bank of protected rivers, with management responsibility given to the nearest federal agency. Rivers were defined according to three categories: "recreational," which allowed for the most development along the shoreline; "scenic," a midlevel category; and "wild," which allowed the least modification. The law also established a process for new river designation, including the requirement of a five-year study before bills could be proposed to Congress. In addition, the Act mandated the creation of cooperative management plans for each designated river.

Ten years later, in 1978, the system had grown to forty-three rivers from the original eight, but advocates had hoped for much more. Critics suggested that this was due in part to the cumbersome designation process and the lack of a single federal management agency for wild rivers. Environmental activists attempted to fill this gap by creating the American Rivers Conservation Council in 1973 (today called simply American Rivers). The group was instrumental in leading campaigns for expanding the system and for drawing public attention to river health issues via their annual "Ten Most Endangered Rivers" list. Passage of the Alaska Lands Act in 1980 significantly expanded the system by thirty-three new rivers, stretching over 3,200 miles. Later that same year, 125 miles of Idaho's Salmon River was added, along with a number of rivers in California. The year 1992 marked another significant expansion of the system, with 520 miles of rivers in Michigan, 85 miles of the Allegheny River in Pennsylvania, and sections of Big Sur, Sespe Creek, and Sisquoc Rivers in California, all receiving protection under the Act.

The last major expansion of the system occurred with passage of the 2009 Omnibus Public Land Act, which declared four hundred miles of new wild and scenic rivers in Idaho and Oregon. As of 2013, over two hundred rivers or river segments in the United States are designated as wild and scenic according to the 1968 Act. Each of the four major federal land agencies (National Park Service, Forest Service, USFWS, and BLM) is involved in providing administrative oversight.

Significantly, river designation under the Wild and Scenic Act does not necessarily exclude use. In fact, designation preserves existing uses. And while it prohibits any impediments to free-flowing waters, such as dams, the requirement of cooperative river management plans ensures that multiple interests are taken into account in striking a balance between resource use and preservation. Moreover, while there may be broad support for protecting the largest and most well known waterways in the United States, drumming up support for the myriad of smaller tributaries and streams flowing *into* these large aquatic arteries can be an extremely complex task. The case of the Klamath River in California underscores all of these points, including the essential need for collaborative approaches to river planning.

Stretching from southern Oregon to the coast of northern California, where it meets the sea, the Klamath River once supported the third largest salmon run in the western United States. It is fed by Klamath Lake in eastern Oregon, and by several other rivers, including the Trinity and Salmon. In 1981, the river received designation under the Wild and Scenic River Act. Over its 286 miles, which includes a number of major forks and tributaries, approximately 12 miles are designated as "wild," 23.5 miles are classified as "scenic," and the remaining 250 miles are categorized as "recreational."[52] The upper half of the reach is managed by the Forest Service and BLM, while the lower half is administered by the State of California, National Park Service, and several Native American tribes.

The river is unique in supporting numerous anadromous fish species (species that spend portions of their lives in saltwater and freshwater environments), including the Northern California Coast Coho salmon, listed as threatened under the Endangered Species Act; Chinook salmon; steelhead trout; and others. The fishery supports a recreational sport fishing industry, a commercial ocean fishing industry, and Native American subsistence needs, including the provision of resources for ceremonial purposes and the preservation of traditional lifestyles. The river also supports irrigated agricultural production inland, hydroelectricity generation, a number of state and national wildlife refuges, and it serves as a wildlife corridor for a wide variety of species. Finally, the twelve-mile "wild" segment of the Klamath provides whitewater rafting opportunities for recreationists.

Dams have been a problem for salmon runs on the Klamath River since their completion in 1918. In an early effort to address the issue, the state of California banned additional dams on some portions of the river in 1924. But it was not enough. By the late 1990s, increased demands on the river from inland farmers and commercial fishing, coupled with severe drought conditions, brought the river to a state of crisis. In 2002, returning adult salmon suffered one of the largest die-offs ever recorded. In 2006, the commercial salmon fishery was closed along seven hundred miles of the West Coast to help restore the population.[53] In 2010, drought conditions forced a reduction in water deliveries to farmers and marked the twenty-fourth year in a row that the Lost River Sucker fishery was closed to indigenous peoples upstream.

In a collaborative attempt to solve this problem, American Rivers and other NGOs began working with the PacifiCorp energy company, federal and state agencies, and representatives from the affected Native American nations to reach a deal whereby five hydroelectric dams might be removed from the Klamath River. A central component of this effort was to show that the cost of dam removal (approximately $90 million) was significantly less than the cost of upgrading the old dams in order to bring them into compliance with current regulatory standards. Activists also underscored the fact that the forgone electricity production amounts to approximately 1 percent of PacifiCorp energy demand in the region. An environmental impact statement produced by the Federal Energy and Regulatory Commission supported these claims, finding that removal and purchase of energy lost by the dams would save customers $28 million per year.

The hope for the Klamath Basin Restoration Agreement (KBRA) is that by removing the dams, the fisheries will be restored to their former vibrancy, allowing water temperatures to cool and for salmon to return to inland headwaters where they have not been seen for almost a century. Meanwhile, upstream water users, including inland farmers and Native groups, may be willing to give up some water claims if in return they receive guarantees of continued water access that lessen uncertainties for agricultural production and subsistence uses in the future. Though the KBRA hit some snags with the emergence of a local Tea Party movement in 2012,[54] supporters remain optimistic that compromise and collaboration will ultimately carry the day and convince a divided Congress to authorize the measure.

National Scenic Trails

On October 2, 1968, the very same day that President Johnson signed the Wild and Scenic Rivers Act, he also signed the National Trails System Act into law. In fact, campaigns for the two bills shared many of the same challenges, as well as many of the same arguments and advocates. Just as wild river proponents could trace the history of their issue back to the early days of U.S. conservation history, so, too, could supporters of a national trails system. In fact, as discussed early in this chapter, those who first championed the idea of national trails were deeply involved with the burgeoning wilderness movement.

Back in 1921, when the idea of wilderness was just beginning to be articulated by the likes of Arthur Carhart and Aldo Leopold, Benton MacKaye published his proposal for the Appalachian Trail.[55] MacKaye, who later served as Wilderness Society president from 1945 to 1950, developed his original vision with the goal of providing sustainable and planned communities for underprivileged families across the spine of the Appalachian Mountains. However, over time, MacKaye increasingly defined his trail project in terms of an interconnected series of wilderness areas—as a "pathway defined by wilderness."[56]

Local and regional interest in his idea led to trail construction by volunteer hiking groups long before federal agencies became involved. This was fitting in a way, since much of the two-thousand-mile trail cut across private property. Groups such as the Appalachian Mountain Club and the Green Mountain Club were instrumental in marking and maintaining the new trail, but they needed to work with private landowners to secure permission. To help broker these agreements and organize the host of volunteers provided by local groups, MacKaye and his supporters formed the Appalachian Trail Conference (ATC) in 1925. Progress remained slow, however, until Arthur Perkins and later, Myron Avery, took the helm of the ATC. Helping to support new local clubs, develop hiking maps and guides, and make arrangements with federal and state officials, the two men provided the necessary leadership to finally complete the Appalachian Trail. By 1937, one could actually hike a continuous trail from Mount Oglethorpe in Georgia to Mount Katahdin in Maine.

One of the biggest obstacles for the Appalachian Trail was the changing ownership of private parcels and road construction activity. In 1938, the ATC began working with the National Park Service and Forest Service to demarcate zones where the trail cut through public lands in order to ensure the primitive quality of the lands on either side. Such arrangements were successful to some degree, but did not stop all projects, including the Blue Ridge Parkway in Shenandoah National Park (discussed earlier). Nonetheless, the ATC continued to broker individual agreements with federal and state authorities with mixed results.

Finally, in the mid-1960s, ATC chairman Stanley Murray met with Wisconsin senator Gaylord Nelson, who became an ardent supporter of the need for national legislation, including a national network of wilderness trails. Like those who supported the campaign for wild rivers, trail advocates benefited greatly from the recently released ORRRC report. By calling for increased outdoor recreation opportunities for all American citizens, the report appeared to directly endorse the idea of a national trail system. With powerful backing from the Johnson administration, Congress soon passed the National Trails System Act of 1968.

The law immediately established two scenic trails, the Appalachian Trail and the Pacific Crest Trail (along the spine of the Sierra and Cascade Ranges, from Canada to Mexico) as part of a new National Trails System. It also identified fourteen additional trails for consideration and established a second category of National Historic Trails that celebrated American heritage, including the Lewis and Clark Trail, the Mormon Pioneer Trail, the Oregon Trail, and the Iditarod in Alaska.

The unique challenge for the Appalachian Trail and subsequent trails was (and remains) the discordant array of land types involved over such vast distances. Like wild and scenic rivers, the paths traverse many different kinds of federal land units (national parks, national forests, national wildlife refuges, BLM lands) as well as state trust lands, municipal areas, and of course, private property. As such, even with congressional statutory status, the day-to-day management of national trails is frequently

Table 8.2.   The National Scenic Trails by Year of Federal Designation

Year	Name	Location	Distance (miles)
1968	Appalachian Trail	Georgia to Maine	2,200
1968	Pacific Crest Trail	California to Washington	2,650
1978	Continental Divide	Montana to New Mexico	3,100
1980	North Country	New York to North Dakota	4,600
1980	Ice Age	Wisconsin	1,000
1983	Florida	Florida	1,400
1983	Natchez Trace	Tennessee to Mississippi	444
1983	Potomac Heritage	Pennsylvania, Virginia, Maryland	830
2009	Arizona	Arizona	800
2009	Pacific Northwest Trail	Montana to Washington	1,200
2009	New England	Connecticut to New Hampshire	220

carried out by local citizen groups that volunteer their time. Such groups serve to maintain the trails, promote them to the general public, and coordinate projects and public events. While federal agencies actively manage trail segments that traverse their jurisdictions, the stewardship of the pathways that lie "between" often falls to these local organizations.

Partly for this reason, even today, only two scenic trails in the National Trail System are completed to the extent that hikers may walk from one end to the other on designated pathways: the Appalachian Trail and the Pacific Crest Trail (PCT). Though both were designated with the passage of the 1968 Act, only the AT was complete at the time. The PCT was a different story. Trail construction began in 1935 under the leadership of the PCT Conference organized by Clinton Clark, Ansel Adams, the Boy Scouts of America, and local YMCA groups. However, the PCT would not reach completion until fifty-eight years later, in 1993. As of 2013, there are eleven national scenic trails in the United States (table 8.2), and nine of them still require significant land purchases and negotiated permissions with private landowners before they can be "through-hiked" by travelers. There are also nineteen national historic trails and thousands of national recreational trails spread throughout the nation.[57]

In 1995, the Partnership for the National Trail System was formed to provide a forum to connect local citizen groups with state and federal agencies in support of trail construction and maintenance. The organization hosts an annual conference to promote collaboration and offers training for local organizations on budget issues, planning, and management.[58] Nonetheless, the future integrity of the national scenic trail system relies heavily on the strength of local organizations to work collaboratively with their neighbors to achieve mutually beneficial solutions to the riddle of interwoven public and private lands that comprise these unique pathways.

9

Parting Thoughts

The mountains are calling and I must go.

—John Muir[1]

Having completed this figurative "through hike" across America's public lands, one may ask, "Where do we go from here?" What lessons can be drawn from this excursion, and what does the future hold? We have explored the public land system from acquisition, to disposal, to conservation, and then considered individual land unit types in terms of defining characteristics, key turning points, and management challenges. Now, it can be helpful to step back and reflect on the major themes that tie this disparate system of lands together into a coherent whole. Below, I offer some parting thoughts on the enduring legacy of the nineteenth-century ideas of nature discussed in the opening chapters, the distinctions and interconnections among different public land types, and finally, the promise of collaborative conservation as America's public land system moves into the twenty-first century.

MAPPING CONCEPTUAL CONTINUITIES

In his influential book, *Crossing the Next Meridian*, law professor Charles Wilkinson argued that twentieth-century environmental politics in the American West remained hobbled by nineteenth-century laws and regulations. Federal laws and administrative rules governing water rights, reclamation, hard rock mining, grazing fees, and timber production—what

he called the "lords of yesterday"—followed a logic of resource development that possibly made sense back in the heyday of early American
industrialization. But in the late twentieth century, these priorities no longer aligned with the values of modern society. Concerned that the public
was largely unaware of these laws and regulatory standards, Wilkinson
sought to bring the "lords of yesterday" to light in the hope of sparking
conversation and possibly stirring up public support for policy reform.[2]

In a similar vein, one purpose of this book has been to pull back the
curtain to expose certain legacies of nineteenth-century thinking that
continue to wield significant influence upon America's public land system, even as we move into the twenty-first century. Though influential,
these conceptions of nature still go largely unnoticed, not only by the
general public but also, at times, by resource managers and many others with an interest in public lands. Despite the "break with the past"
assumption embedded within dominant narratives of conservation in
the United States, the ideas of nature-as-commodity and as something
that is unpeopled, eternal, and pristine, have continued to shape many
of the most challenging policy and management issues facing America's
public land system today.

At times these notions are blatantly and explicitly conveyed in formal
statutes, such as the 1872 Hard Rock Mining Law. But as we have seen,
more often than not, they are expressed in much more subtle ways. They
remain on the margins of public awareness, lurking in complex regulatory histories and deeply woven into the very laws and institutions presumably (ostensibly?) created to guard against them. The power of these
ideas derives in no small part from the stealthy manner in which they
exert themselves. It has been said that fish will be the last creatures to
discover the idea of water. It is, therefore, no surprise that the things we
take for granted are often the most difficult for us to acknowledge and
understand. And for that reason, they can exert the greatest influence on
the way we see the world. Hidden in plain sight, these ideas can shape
our thinking without our ever knowing.

In the case of America's public lands, we sense the influence of these
ideas in conflicts over snowmobiles in Yellowstone, grazing management
in Grand Staircase-Escalante National Monument, the challenges facing
the National Elk Refuge, or in recent debates over the fate of Tongass
National Forest and the Roadless Conservation Rule. And once acknowledged, these nineteenth-century conceptions of nature can explain much.
For one thing, they offer resolution to an apparent contradiction: How is it
that a country so committed to private property and the principles of free-
market commodity production can be willing to set aside one-third of its
land area under "public ownership," and indeed, be the first modern state
in the world to do so? The answer, at least in part, is that the presumed

"break with the past" triggered by the nineteenth-century environmental conservation movement never really happened.

At least in practice, and possibly for the sake of political expediency, early activists allowed a number of "conceptual stowaways" into the conservation movement: ideas that jumped from one ideological train car to the next precisely at the moment of decoupling. As a consequence, the concept of nature-as-commodity became woven into the very rationale offered to create the world's first national park. Just as politicians passionately argued for Yellowstone's protection on the basis that it held no desirable economic value, the park's most powerful backers—the railroads—clearly held the opposite view. Despite political rhetoric to the contrary, many of Yellowstone's congressional supporters valued it precisely in economic terms. Defined as nature tourism, the idea of the national parks was infused with nature-as-commodity thinking from the very beginning.

The concept of nature-as-commodity can be traced out in other public land types as well. The Forest Service embraced this perspective as part of its Progressive Era conservation approach to resource management. National forests were set aside to allow for a more rational science-based approach to timber management. But at the same time, commercial development remained a central goal. In a similar fashion, recreational tourism found a constituency in the sport hunting groups advocating for national wildlife refuges, not to mention the backpackers and outdoor enthusiasts supporting the modern wilderness, wild rivers, and national trails movements. And of course, early political support for the creation of the Bureau of Land Management in the 1940s was premised on the notion that the new agency might better facilitate commercial resource development and exploitation of the public domain than the Forest Service.

Similarly, the idea of nature as unpeopled, eternal, and pristine exerted tremendous influence in its own right. By adopting a static view of nature, policymakers and resource managers could argue that simply erecting and policing boundaries around parks, forests, and refuges was sufficient to preserve wildlife and landscapes forever in their pristine state. This is illustrated in early approaches to wildlife management in the national parks and wildlife refuges, the adoption of fire-suppression policies on national forests, and, for at least some advocates, the primary rationale for establishing a wilderness preservation system.

But perhaps most importantly, these conceptual continuities also explain why many modern-day conflicts in public land management and policy are so frequently deemed intractable. As documented throughout this book, some of the most difficult public land use challenges stem directly from the continued presence of nature-as-commodity thinking, as well as the notion of nature as unpeopled, eternal, and pristine within the very same laws and institutions established to champion nature protection according to decidedly *non*-commodity-based ideals.[3]

DIVERSITY WITHIN THE PUBLIC LAND SYSTEM

Another goal has been to make some sense out of the differences and distinctions among various types of units in America's public land system. As noted earlier, this is necessary, in part, to avoid conflation of the entire system with the national parks. But more importantly, it is also premised on the notion that one cannot fully understand one type of public land without acknowledging its relationship to the others. Recognizing these interconnections is critical given the increasing number of ecosystem and region-based initiatives that necessitate coordinated management across institutional boundaries. These relationships are expressed in a number of ways, including management priorities and the institutional evolution of various public land agencies.

On a continuum of resource management priorities ranging from nature preservation to commercial resource development, most would place wilderness areas and the BLM lands on opposite ends of the spectrum, with the national parks, forests, and wildlife refuges lying somewhere in between. Nonetheless, the competitive coevolution of the national parks and national forests offers perhaps the starkest contrast between federal land types in the system.

The national parks represent the earliest designated type of federal lands to be set aside for protection, while the Forest Service stands as the oldest federal land management agency.[4] But while they both represent a land retention alternative to unfettered nineteenth-century commercial resource development, they do so in distinctive ways. The parks focus on environmental preservation and recreation. In contrast, the national forests adhere to a progressive conservation philosophy that promotes resource development via the application of scientific management. The fact that the two agencies reside in different branches of the federal government (the Departments of the Interior and Agriculture, respectively) underscores this difference, and it stems in part from Gifford Pinchot's effort to distance the Forest Service from the Romantic philosophy of nature preservation, as evidenced in the Hetch Hetchy conflict.

Yet inasmuch as the two agencies have sought to highlight their differences, they have also heavily influenced and shaped one another in a myriad of ways through their mutual competition for limited land and government resources. Frequently, this competition has devolved into a race to broaden or redefine managerial priorities and institutional mandates. At the turn of the nineteenth century, new national parks were frequently carved out of national forestland on the rationale that the Park Service could better preserve the lands for recreational purposes. To stop the hemorrhaging of national forest lands to the park system, the Forest Service vigorously opposed the creation of the National Park Service in 1916, claiming that the new agency would be redundant. Did not the Forest Service already manage lands for recreational purposes as part of its multiple-use mandate? Though its effort to block the National Park Service Act ultimately failed—falling before the juggernaut that was Stephen Mather—it spurred the Forest Service to begin building campgrounds, leasing summer cabin sites, and promoting recreation as a valid use of national forest resources.

Moreover, the Park Service's aggressive promotion of motorized tourism triggered not only Forest Service support for the creation of "primitive areas" in the 1920s but also provided the impetus for the 1935 creation of the Wilderness Society, leading to the passage of the Wilderness Act in 1964. In fact, one cannot fully understand the emergence of the wilderness movement—including the creation of the world's first wilderness area on the Gila National Forest in 1924—without acknowledging the role of national park policy toward automobile tourism in the 1920s and 1930s. In short, some of the very characteristics that distinguish national parks, national forests, and wilderness areas result directly from their close historical relationship to one another.

In contrast to the national parks and forests, the national wildlife refuges and BLM lands share a very different set of defining characteristics. First of all, each of the managing agencies, the Bureau of Land Management (BLM) and U.S. Fish and Wildlife Service (USFWS), respectively, is located in the Department of the Interior, like the Park Service. But despite this fact, they both adhere to the multiple-use philosophy pioneered by the Forest Service rather than the preservation/recreation mandate governing the national parks. Moreover, both BLM lands and national wildlife refuges are further differentiated from the national parks *and* national forests by their need to grapple with a regional bias, insofar as the overwhelming majority of their land holdings lie in the westernmost states.

But perhaps their most important defining characteristic derives from the unique institutional evolution of their managing federal agencies. Though both agencies came into being in the 1930s and 1940s as products of New Deal initiatives, the BLM and USFWS have struggled to separate themselves from nineteenth-century predecessors. Unlike the Forest Service and Park Service, which were created largely from scratch,[5] both the USFWS and BLM evolved from institutions with decidedly checkered pasts. Consider the case of the USFWS. Its institutional genealogy included the Bureau of Biological Survey and the Bureau of Fisheries. The former focused largely on the extirpation of undesirable species to benefit agricultural interests, while the latter was crippled through most of its history by a dependency upon and service to the interests of the commercial fishing industry. Similarly, the BLM emerged from two separate entities: the General Lands Office, whose primary function was to oversee the disposal of the public domain; and the ineffective U.S. Grazing Service, an agency described by critics as thoroughly "captured" by western livestock interests.

These troubling institutional histories have continued to yield effects. In the constant struggle to escape the long shadows of their respective pasts, both the BLM and USFWS have grappled with the influence of various industrial interests. Just as the Forest Service has fought to distance itself from the commercial timber industry at various points in its past, the USFWS has faced a similar task with the sport hunting industry. Meanwhile, the BLM has long contended with mining and livestock interests. However, even on the issue of interest group influence, relationships between the different agencies have mattered. As noted above, the perception among western livestock interests that the Forest Service would be more difficult to influence on public grazing issues than a new federal agency helped provide the necessary political support for the creation of the BLM in the first place. In short, understanding the relations between various types of public lands can not only help to bring their differences into sharp relief, but in some cases, explain key characteristics as direct outcomes of agency interaction and competition.   ·

THE PROMISE OF COLLABORATIVE CONSERVATION

Finally, as we look ahead, what lies on the horizon for America's public lands? I offer observations on three still-unfolding developments, all of which suggest an important place for collaborative approaches in future public land management and policy discussions. The first derives from the observation that the era of new, large-scale, land-based declarations of national parks, forests, and wildlife refuges is long since over.[6] The occasional designation of wilderness areas and national monuments notwithstanding (e.g., President G. W. Bush's national marine monuments in the Pacific or President Clinton's declaration of the Grand Staircase-Escalante in southern Utah), on the whole, broad, regional landscapes currently considered as candidates for public protection are either already managed via a mix of municipal, state, and federal agencies or lie in private hands. Hence, the 2012 "designation" of the new 150,000-acre national wildlife refuge in the Everglades Headwaters (chapter 6) is really a declaration of *intent* by the federal government to work with private landowners and land conservancy organizations to broker easements that will allow for varying measures of environmental protection over a vast landscape. This is in keeping with other regional-scale efforts to protect watersheds and ecosystems, including the Chesapeake Bay, the Greater Yellowstone Ecosystem, the Klamath River Basin, and the Santa Monica Mountains in California. All of these efforts are comprised of a diverse array of land designations governed by a similarly disparate mix of government agencies, NGOs, and private landowners operating at different jurisdictional scales, from the national to the local. Needless to say, coordinating such efforts demands a collaborative approach to environmental management in order to work effectively with the multiplicity of players and stakeholders involved.

The second observation is largely a consequence of the first and specifically concerns the relationship between public and private space. By actively seeking to incorporate privately owned properties, in the form of conservation easements, within the rich and diverse tapestry of land types comprising regional and watershed-based conservation schemes, federal managers are granting private landowners a much greater role in public land management and policy decisions. But even more significantly, the concept of ecosystem management that underlies these schemes demands that attention be paid to the human communities that reside within and adjacent to protected landscapes. In each of these ways, federal land managers are increasingly and necessarily blurring the notion of a strict public-private divide, in both conceptual and actual material "on the ground" terms.

In practice, the idea of ecosystem management acknowledges the need for "working landscapes" and is committed to finding a space for land use and livelihood activities as part of a more sustainable form of resource development. By breaking down rigid divisions between a "pure" nature devoid of humans on the one hand, and places where humans reside on the other, managers are pulling questions of zoning, regional planning, and resource development to the center of discussions over environmental conservation. At a conceptual level, these ideas clearly move us away from the notion of nature as unpeopled, eternal, and pristine, ideas that for decades have provided foundational rationales for the creation of public lands, public land agencies, policies, and management philosophies.

Efforts to integrate human communities within conservation areas have a long history in protected areas in the Global South (albeit with varying degrees of accommodation for indigenous peoples). And they found early expression in the United States in Alaska under ANILCA. But increasingly, these approaches are emerging as the new norm in conservation, requiring a prominent role for collaboration as a means for convening diverse stakeholder discussions and working out potentially sustainable solutions to land management problems.

A third and final observation pertains to the growing effects of climate change and the challenges it poses to America's public land system. Already, increases in global average surface temperatures are rendering demonstrative environmental impacts, most dramatically documented in the receding ice flows in Alaska's Glacier Bay and in Glacier National Park in Montana. As noted earlier, the National Park Service predicts that glaciers will disappear entirely from Glacier National Park by the year 2030.[7] Can there be any more explicit evidence of the realities of climate change? Species change induced by climatic alterations is also under way. Aside from the media-drenched plight of polar bears, a multitude of native species uniquely adapted to a wide variety of vulnerable environments are quietly showing signs of stress on their populations (e.g., the American pika, desert tortoise, and many others).[8]

In the face of such rapid environmental change, the fallacy of an eternal, pristine, and therefore, static nature quickly melts away (pun intended). As noted earlier, the danger is very real that the landscapes or species for which public lands were initially set aside for protection may no longer align with the rigid boundaries delimited by national parks, forests, or wildlife refuges. Under conditions of climate change, the dynamism of natural systems, which has always existed, is speeding up. Wildlife species may migrate to northern latitudes or other locales, or simply die out. The vegetative communities that comprise wildlife habitats will also migrate, albeit at much slower rates, thus creating mismatches between species and habitats that benefit the "generalist" species most able to adapt. Hence, the unique landforms for which some national parks were established to ensure their protection may find themselves transformed into something quite different. Already, some forested watersheds around which national forests were established in order to provide careful management are experiencing changes in species composition and may witness unexpected changes in their ability to provide water resources on a seasonal basis. And so on.

All three of these future trends—mixed land unit types comprising ecosystem-scale conservation initiatives, the accommodation of human use and habitation within protected public/private landscapes, and the need to grapple with the increasingly acute impacts of climate change— point to the porosity and complexity of public land issues. As a consequence, unilateral decisions by federal agencies applied to large-scale conservation landscapes are becoming increasingly rare as managers turn to collaborative approaches. At its best, collaborative conservation can offer a clear way forward for managers grappling with multiple jurisdictions across a common landscape. It can provide a means for cultivating common ground among diverse interests, for incorporating the needs, concerns, and knowledge of local residents in governance processes, and for discovering innovative solutions to the unprecedented challenges presented by global climate change.

Of course, at its worst, collaboration may accomplish none of these things. It may devolve into an exercise of exclusion and parochialism, using the shield of "community participation" to justify management goals quite detached from the principles of environmental science, open participation, biodiversity protection, and sustainable resource development. At the end of the day, collaborative conservation is only a tool, one that requires careful application in order to fulfill its potential.

Nor is it the only tool. Congress will continue to set policy via the legislative process, presidents will draft executive orders, courts will issue rulings, and federal managers will craft management plans and carry out day-to-day land use decisions. But amid this complex array of mechanisms, for the reasons cited above, collaborative conservation is uniquely situated to make a positive contribution to the task of finding policy and management solutions for the challenges facing America's public land system in the twenty-first century.

The next time you visit a national park, camp in a national forest, ride a wild river, or hike a national scenic trail, perhaps some of these ideas will come to mind. Aldo Leopold wrote, "We can be ethical only in relation to something we can see, feel, understand, love or otherwise have faith in."[9] Hopefully, this book provides a measure of understanding that helps us advance a few more steps on the journey toward an ethical treatment of nature in the guise of public lands. If so, perhaps we will be better able to pass on this unique heritage to the next generation in a healthier, more robust, and diverse condition than we found it.

Appendix A

Major U.S. Public Land Laws and Other Key Turning Points

Year	Law	Description
1841	Preemption Act	Allowed squatters on the public domain to purchase claimed land for a minimal price without going through public auction.
1862	Homestead Act	Intended to encourage western settlement. The law gave 160-acre land claims to settlers for a $26 fee if they worked the land for five years. Land could also be purchased after six months for $1.25 per acre.
1862	Morrill Act	Provided land to states to finance institutions of higher education.
1862–1870	Federal Railroad Grants	During this period, over ninety-four million acres of public domain land was given to the railroad companies to help subsidize the construction of new rail lines.
1872	General Mining Law	Still in effect today, this law defines the process for staking valuable hard rock mining claims on the public domain. It includes several highly controversial practices, including the patenting of claims as private land parcels, thereby creating inholdings within public land units.
1872	Yellowstone National Park Act	Established the world's first national park, located in Wyoming, Montana, and Idaho.
1873	Timber Culture Act	An attempt to encourage settlement in arid conditions, the law allowed homesteaders with 160-acre claims to receive an additional 160 acres if they promised to plant trees on forty acres.

Year	Law	Description
1876	Desert Land Act	Another law designed to facilitate settlement in the arid West. It allowed settlers to claim 640-acre homesteads if they promised to irrigate the land within three years.
1878	John Wesley Powell publishes *Report on the Lands of the Arid Region of the U.S.*	Provided insights on how to best adapt human settlement to arid environments.
1890	Creation of Yosemite, Sequoia, and General Grant National Parks	The next set of national parks established after Yellowstone, due largely to campaigns led by John Muir.
1891	General Revision Act	Repealed the Timber Culture Act, revised the Desert Land Act, and included an amendment granting power to the president to set aside portions of the public domain as national forest reserves.
1892	Sierra Club established	Led by John Muir, the Sierra Club is one of the oldest and most influential environmental organizations in the United States.
1894	Yellowstone Wildlife Protection Act	The first federal wildlife protection law. Prohibited hunting and the removal of wildlife from Yellowstone National Park, making Yellowstone the nation's first wildlife preserve.
1897	Forest Organic Act	Established the rationale for national forest reserves: to provide a continuous supply of timber and provide water resources for the American people.
1900	Lacey Act	Established federal authority to prohibit the transport of wildlife killed in violation of state laws across state lines. Also gave the Secretary of Agriculture the power to restore, introduce, distribute, and preserve game birds in accordance with state laws.
1902	Reclamation Act	Funded irrigation projects on arid public lands in the western United States and established the U.S. Reclamation Service (later renamed the Bureau of Reclamation) to carry them out.
1903	Pelican Island Bird Reservation, Florida	The nation's first wildlife preserve established via executive order, by President Theodore Roosevelt.
1905	Reorganization Act	Moved the national forest reserves from the Interior Department to the Agriculture Department. Formally established the Forest Service as the managing agency under the leadership of Gifford Pinchot, first chief forester of the United States.

Year	Law	Description
1906	Antiquities Act	Granted authority to the president of the United States to create national monuments via executive order.
1911	Weeks Act	Allowed the expansion of the national forest system in the eastern United States and underscored fire suppression as a primary management goal.
1916	National Park Service Act	Established the National Park Service and defined the purpose of the park system: to "conserve the scenery" and promote recreation in such a way as to "leave them unimpaired for future generations."
1920	Mineral Leasing Act	Set up a new regulatory structure, including the requirement of royalties paid to the federal government, for coal, oil, natural gas, and other hydrocarbon resources in the public domain. Previously, these minerals were governed under the 1872 General Mining Law. The law also applied to phosphates, potassium, sodium, and sulphur resources.
1924	Creation of the first "Primitive Area" on the Gila National Forest in New Mexico	Considered the world's first wilderness area. It was due largely to the efforts of Aldo Leopold.
1934	Duck Stamp Act	Also known as the Migratory Bird Hunting and Conservation Act. It provided funding for the creation of new wildlife refuges identified under the 1929 Migratory Bird Act with proceeds from the sale of hunting permits in the form of adhesive stamps.
1934	Taylor Grazing Act	Established grazing districts on the public domain and set up a permit and fee system for those wishing to graze livestock.
1935	Wilderness Society established	Created by Bob Marshall, Aldo Leopold, Benton MacKaye, and others in support of wilderness preservation.
1936	National Wildlife Federation established	The nation's largest organization explicitly devoted to the conservation of all wildlife.
1946	Reorganization Plan	Issued by President Truman to merge the General Land Office and Grazing Service to create the new Bureau of Land Management.
1946	First Sagebrush Rebellion	Failed congressional effort to transfer the public lands to the states, where they would then be auctioned off to private landholders.

Year	Law	Description
1960	Multiple Use and Sustained Yield Act	Identified five major uses of national forests: range, timber, watershed protection, wildlife and fish habitat, and outdoor recreation (including wilderness). It also prohibited the Forest Service from logging more wood than could be grown back in a given year.
1964	Classification and Multiple Use Act	Mandated the BLM to review its public land holdings and determine which should be classified according to a variety of multiple-use purposes and which should be "disposed" of or sold off into private ownership.
1964	The Wilderness Act	Established the National Wilderness Preservation System, defined the concept of wilderness, and detailed the process for federal wilderness designation.
1965	Land and Water Conservation Act	Funding to purchase lands to expand the national park, forest, or wilderness system with monies provided from offshore oil drilling fees. However, Congress regularly diverts these funds for other purposes.
1966	National Wildlife Refuge System Administration Act	Established the national wildlife refuge system in the United States, bringing together various fish, game, and bird reserves, along with fish hatcheries and "waterfowl production areas" under the authority of one agency, the U.S. Fish and Wildlife Service.
1968	National Wild and Scenic Rivers Act	Authorized Congress to set aside and protect sections of free-flowing rivers in the United States according to a three-tiered classification scheme.
1968	National Trails System Act	Authorized Congress to establish a system of national trails according to four categories: national scenic trails, national historic trails, national recreation trails, and "connecting" trails that provide access to or connect the trail types noted above.
1968	Pryor Mountain Wild Horse Range	Created by Interior Secretary Stewart Udall as the first wild horse range in the United States. Managed by the BLM.
1969	National Environmental Policy Act	Established a precautionary approach to public land management by requiring the creation of Environmental Impact Statements (EIS) for all federal resource management projects. Also mandated public participation in their creation.
1971	Wild and Free Roaming Horse and Burro Act	Provided federal protection for wild horses and burros that inhabit federally owned lands.

Year	Law	Description
1972	Federal Advisory Committee Act	Set limits and regulations regarding the role of advisory boards, committees, and other groups (including collaborative groups) that may offer advice to federal managers. Designed to shield agency decision making from the undue influence of special interests.
1973	Endangered Species Act	Provides federal protection for species and habitats designated as "threatened" or "endangered."
1971–1973	Roadless Area Review and Evaluation I (RARE I)	The Forest Service's first attempt to inventory national forests for potential wilderness designations that ended in a number of lawsuits.
1975	Eastern Wilderness Act	Established sixteen new wilderness areas in national forestlands in the eastern U.S. and directed the Forest Service to drop its "pure" definition of wilderness in favor of a much broader term.
1976	Roadless Area Review and Evaluation II (RARE II)	Second attempt by the Forest Service to review its holdings for potential wilderness areas. As before, the results were met with lawsuits. Ultimately, the effort led to a series of state-based wilderness designations.
1976	National Forest Management Act	Established strict limitations on the use of clear-cut harvesting methods. Required the creation of long-term comprehensive management plans for each national forest.
1978	Federal Land Policy and Management Act	The "organic act" for the Bureau of Land Management. Effectively repealed all Homestead Acts in all states excluding Alaska. Required the BLM to manage according to a multiple-use approach, to develop comprehensive management plans, and to inventory its lands for potential wilderness designations.
1978	Endangered American Wilderness Act	Protected a total of 1.3 million acres by establishing thirteen new wilderness areas and expanded four existing wilderness areas.
1979–1982	Second Sagebrush Rebellion	In response to FLPMA, changes in grazing fees, and efforts to reinforce conservation measures on public lands, western congressional delegations once again tried and failed to transfer control of federal lands and resources to the states.

Year	Law	Description
1980	Alaska National Interest Lands and Conservation Act	Expanded the national park system, national forest system, national wildlife refuge system, and national wilderness preservation system by nearly eighty million acres.
1994	Northwest Timber Plan	Forest planning effort to resolve the matter of protecting habitat for the northern spotted owl. One of the first plans to use ecosystem-based and collaborative approaches to national forest management.
1994	California Desert Lands Act	Established 7.7 million acres of wilderness in California.
1996	Grand Staircase-Escalante National Monument	President Clinton declares the first national monument to be managed solely by the BLM.
1997	National Wildlife Refuge System Improvement Act	The "organic act" for the U.S. Fish and Wildlife Service, identifying wildlife conservation and preserving biological integrity as primary agency goals. It also defined appropriate recreational activities in the refuge system and required comprehensive management plans.
1998	National Wildlife Refuge System Volunteer and Community Partnership Enhancement Act	To encourage public participation in USFWS programs, build public awareness, and promote public education of the national refuge system.
2000	National Landscape Conservation System	Created by Secretary of the Interior Bruce Babbitt to manage all special conservation areas in the BLM together as part of a single system.
2001	Roadless Area Conservation Rule	Rule issued by the Forest Service to prohibit logging in any national forest area devoid of roads. Protected some fifty-six million acres from logging.
2003	Healthy Forest Restoration Act	Allowed expedited timber harvests on national forests and BLM lands to reduce threats from wildfire, insect outbreaks, and disease.
2005	Energy Policy Act	Directed public land agencies to prioritize the production of fossil-fuel-based energy resources.
2009	Public Land Omnibus Act	Protects over two million acres of wilderness in nine states, one thousand miles of wild and scenic rivers, three national parks, three conservation areas, four national trails, ten national heritage areas, and one national monument.

Appendix B

Units within the National Park System

Unit Type	No.	Description
National Parks	59	Established by Congress, parks generally provide the highest level of protection of historical or environmental resources within the national park system. They often comprise relatively large areas of significant scenic, scientific, or natural value. Hunting, mining, logging, and other extractive activities are usually prohibited.
National Monuments	78	Established by the president via executive order under the Antiquities Act of 1906. Levels of protection vary by monument. Many were later redesignated by Congress as national parks.
National Preserves	18	Established by Congress to protect certain resources, but it may allow uses that are otherwise prohibited in national parks, monuments, or other unit types.
National Reserves	2	Includes both public and privately owned lands as part of a larger management area designed to protect certain resource values.
National Lakeshores	4	Protects natural values while providing public access and opportunities for water-oriented recreation. All existing units are located on the shores of the Great Lakes.
National Seashores	10	Protects natural values while providing public access and opportunities for water-oriented recreation on the nation's beaches and coastlines.
National Recreation Areas	18	Areas set aside primarily for the opportunities they provide for outdoor recreation.

271

Unit Type	No.	Description
National Rivers, National Wild and Scenic Rivers	5 10	Stretches of free-flowing rivers given various levels of protection according to their classification as "recreational," "scenic," or "wild."
National Parkways	4	Scenic or historic ribbons of land flanking roadways that offer documentation to visitors on the history or natural beauty of the area.
National Scenic Trails	3	Long-distance footpaths that traverse areas of special scenic, scientific, or historical interest. They often extend well beyond national park system lands.
National Battlefields, National Battlefield Parks, National Battlefield Sites, National Military Parks	11 4 1 9	Established by Congress to protect areas associated with American military history. Sites are generally smaller than parks; however, the NPS does not distinguish between them in terms of managerial priorities or approach.
National Historic Parks	46	Areas of national historical interest set aside for preservation purposes. Parks are generally larger and/or more complex than historic sites.
National Historic Sites	78	Areas of national historical interest set aside for preservation purposes.
National Memorials	29	Areas of national historic interest that are primarily commemorative in nature. They need not be directly linked with the history of their subject, such as the Lincoln Memorial in Washington, D.C.
Other Units	—	These include places such as the National Mall in Washington, D.C. The NPS also administers a great number of wilderness areas (chapter 8) as part of the agency's stewardship duties.

Note: For a complete listing of national park system units by type, see National Park Service, "Nomenclature of Park System Areas," http://www.nps.gov/history/history/hisnps/NPSHistory/nomenclature.html.

Notes

INTRODUCTION: WHY PUBLIC LANDS?

1. The number of acres managed alone or in part by the U.S. Fish and Wildlife Service grows to over three hundred million acres if one includes the national monuments and marine sanctuaries in the South Pacific (discussed in chapter 6).

2. William Cronon, "The Trouble with Wilderness," in *Uncommon Ground: Rethinking the Human Place in Nature*, ed. William Cronon (New York: W. W. Norton, 1996), 69–90.

3. For an in-depth discussion of this issue, see Mark D. Spence, *Dispossessing the Wilderness: Indian Removal and the Making of the National Parks* (New York: Oxford University Press, 1999).

4. Dyan Zaslowsky and T. H. Watkins, *These American Lands: Parks, Wilderness, and the Public Lands* (Washington, DC: Wilderness Society and Island Press, 1994).

5. Prior to Zaslowsky and Watkins, one of the best-known works on the history of America's public land laws and policies derived from a congressional report written for the Public Land Law Review Commission. I refer to the seminal work by Paul W. Gates, *History of Public Land Law Development* (Washington, DC: GPO, 1968).

6. This literature examines management issues pertaining to all different types of lands set aside for conservation purposes, both in the United States and abroad. For example, see Tony Prato and Dan Fagre, *National Parks and Protected Areas: Approaches for Balancing Social, Economic and Ecological Values* (New York: Wiley-Blackwell, 2005) or John Sheail, *Nature's Spectacle: The World's First National Parks and Protected Places* (Washington, DC: Earthscan, 2010).

7. There are too many works to cite here, but notable examples include Roderick Nash, *Wilderness and the American Mind*, 3rd ed. (New Haven: Yale University Press, 1982); John Opie, *Nature's Nation: An Environmental History of the United States* (Fort Worth: Harcourt Brace College Publishers, 1998); and Richard N. L. Andrews, *Managing the Environment, Managing Ourselves: A History of American Environmental Policy* (New Haven: Yale University Press, 1999). In addition, some very useful sources address public lands issues within the context of regional histories or geographies. See, for example, Richard White, *"It's Your Misfortune and None of My Own": A New History of the American West* (Norman: University of Oklahoma Press, 1991), or Gundars Rudzitis, *Wilderness and the Changing American West* (New York: Wiley, 1996).

8. An excellent overview of the field of political ecology can be found in Paul Robbins, *Political Ecology: A Critical Introduction*, 2nd ed. (New York: Wiley-Blackwell, 2011).

9. In 2008, Interior Secretary Dirk Kempthorne ordered that heretofore the Bureau of Land Management lands shall be referred to as the National System of Public Lands.

CHAPTER 1: BUILDING THE NATIONAL COMMONS

1. Rutherford H. Platt, *Land Use and Society: Geography, Law and Public Policy* (Washington, DC: Island Press, 1996).

2. Ira G. Clark, *Water in New Mexico: A History of Its Management and Use* (Albuquerque: University of New Mexico Press, 1987).

3. The thirteen colonies include (in alphabetical order): Connecticut, Delaware, Georgia, Maryland, Massachusetts, New Hampshire, New Jersey, New York, North Carolina, South Carolina, Pennsylvania, Rhode Island, and Virginia.

4. Paul W. Gates, *History of Public Land Law Development* (Washington, DC: GPO, 1968).

5. See Richard N. L. Andrews, *Managing the Environment, Managing Ourselves: A History of American Environmental Policy* (New Haven: Yale University Press, 1999).

6. Gates, *History of Public Land Law Development*.

7. Andrews, *Managing the Environment, Managing Ourselves*, 74.

8. Most accounts of U.S. public lands limit their discussion of the federal domain to continental land acquisitions. I choose to include these "noncontinental" lands, however, because some of them contain territorial or marine resources that constitute significant portions of the modern public land system, either as national parks, monuments, or national marine sanctuaries.

9. Officially known as U.S. Minor Outlying Islands, these include Wake Island, Johnston Atoll, Midway Atoll, Kingman Reef, Palmyra Atoll, Jarvis Island, Baker Island, and Howland Island. In the Caribbean Sea, they include the disputed Navassa Island (claimed by Haiti) and Bajo Nuevo Bank and Serranilla Bank (claimed by Colombia, Nicaragua, and Jamaica, respectively). All islands except Wake, Johnston Atoll, Bajo Nuevo, and Serranilla Bank contain U.S. National Wildlife Refuges.

10. Andrews, *Managing the Environment, Managing Ourselves.*

11. Richard White, *"It's Your Misfortune and None of My Own": A New History of the American West* (Norman: University of Oklahoma Press, 1991).

12. U.S. Department of the Interior, Bureau of Indian Affairs (BIA), "Who We Are," http://www.bia.gov.

13. Ibid.

14. For similar reasons, I also exclude federal military lands from my account of the federal domain. Like the Federal Indian Reserved Lands, U.S. Military Lands differ profoundly from other land types in the public domain in terms of their purpose (in this case, national security) and managerial authority (e.g., the Department of Defense).

15. See R. McGreggor Cawley, *Federal Land, Western Anger: The Sagebrush Rebellion and Environmental Politics* (Lawrence: University Press of Kansas, 1993) and William Chaloupka, "The County Supremacy and Militia Movements: Federalism as an Issue on the Radical Right," *Publius* 26, no. 3 (1996): 161–75.

16. As detailed in later chapters, there are cases in which privately owned properties were gifted to, or purchased by, federal, state, or local authorities in order to create federal land units. However, these actions were done in accordance with constitutional law. The majority of land in the federal system derived from lands already existing in the public domain. To date, legal efforts to challenge federal authority have fallen short.

17. Collaborative conservation is discussed in greater detail in chapter 3. However, excellent accounts of the rise of collaboration and the challenges it faces include Robert Keiter, *Keeping Faith with Nature: Ecosystems, Democracy and America's Public Lands* (New Haven: Yale University Press, 2003) and Edward P. Weber, *Bringing Society Back In: Grassroots Ecosystem Management, Accountability and Sustainable Communities* (Cambridge, MA: MIT Press, 2003). See also Randall K. Wilson, "Community-Based Management and National Forests in the Western United States: Five Challenges," *Policy Matters* 12 (September 2003): 216–24.

18. See William Cronon, ed., *Uncommon Ground: Rethinking the Human Place in Nature* (New York: W. W. Norton, 1996) and Mark David Spence, *Dispossessing the Wilderness: Indian Removal and the Making of the National Parks* (New York: Oxford University Press, 1999).

CHAPTER 2: DISPOSING THE PUBLIC DOMAIN: FROM COMMONS TO COMMODITY

1. Thomas Jefferson, *Notes on the State of Virginia*, ed. William Peden (Chapel Hill: University of North Carolina Press, 1982), 164–5. First published in 1785 (France) and 1787 (United States).

2. William Gilpin, "Address to the U.S. Senate, March 2, 1846," in *Mission of the North American People, Geographical, Social and Political* (Philadelphia: J. B. Lippincott, 1873), 124.

3. Richard White, *"It's Your Misfortune and None of My Own": A New History of the American West* (Norman: University of Oklahoma Press, 1991), 138. For the

definitive history of federal efforts to dispose of the public domain during the first decades of American independence, see Malcolm J. Rohrbough, *Land Office Business: The Settlement and Administration of American Public Lands, 1789–1837* (New York: Oxford University Press, 1968).

4. U.S. Department of the Interior, Bureau of Land Management, *Public Land Statistics 2011, vol. 196* (Washington, DC: GPO, 2012).

5. We will revisit these ideas in chapter 3 in the context of community-based collaborative conservation. Former mayor of Missoula, Montana, Daniel Kemmis, has written extensively on how Jefferson's ideas relate to the governance of federal lands in the American West. See Daniel Kemmis, *Community and the Politics of Place* (Norman: University of Oklahoma Press, 1990) and *This Sovereign Land: A New Vision for Governing the West* (Washington, DC: Island Press, 2001).

6. Rutherford Platt, *Land Use and Society: Geography, Law and Public Policy* (Washington, DC: Island Press, 1996).

7. Richard N. L. Andrews, *Managing the Environment, Managing Ourselves: A History of American Environmental Policy* (New Haven: Yale University Press, 1999), 86.

8. For a detailed account of the General Land Office, again see Rohrbough, *Land Office Business.*

9. White, *"It's Your Misfortune and None of My Own,"* 139.

10. Paul W. Gates, *History of Public Land Law Development* (Washington, DC: GPO, 1968).

11. U.S. Department of the Interior, Bureau of Land Management, *Public Land Statistics 2011.* Eventually, the various Homestead Acts were repealed through a number of subsequent public land laws, including the 1934 Taylor Grazing Act and the 1976 Federal Land Policy and Management Act.

12. Some fifty years later, with passage of the Smith-Lever Act of 1914, Congress would formally establish a system of agricultural extension agencies. These were designed to work with the land grant universities to conduct research and disseminate the latest agricultural knowledge and technologies to farmers.

13. White, *"It's Your Misfortune and None of My Own,"* 146.

14. Ibid.

15. Ibid.

16. Gates, *History of Public Land Law Development.*

17. Ibid.

18. Frederick Jackson Turner, "The Significance of the Frontier in American History," Chicago World's Fair, Chicago, July 12, 1893, American Historical Association, http://www.historians.org/about-aha-and-membership/aha-history-and-archives/archives/the-significance-of-the-frontier-in-american-history.

CHAPTER 3: A PUBLIC LAND SYSTEM EMERGES

1. James Fenimore Cooper, *The Prairie* in *The Works of James Fenimore Cooper*, vol. 2 (New York: P. F. Collier, 1891), 256. First published in 1827.

2. It is important to note that although Yellowstone was the first national park, it did not mark the first federal effort to set aside public lands for environmental

protection. As will be explained later, there are numerous examples of federal efforts to protect lands or resources prior to 1872, including stands of timber for ship masts and deposits of minerals during colonial times. Even unique landscapes, such as the Arkansas Hot Springs, were set aside by Congress in 1832. The 1864 protection of Yosemite Valley also precedes Yellowstone by some eight years. However, in each of these cases, the lands in question did not become *national* parks until decades later. In contrast, Yellowstone was the first national park to receive permanent and continuous federal protection in the United States.

3. A full and detailed accounting of these events is well beyond the scope of this book. The "tragedies" correspond heavily with the westward expansion of the United States and the rise of modern industrialization in the nineteenth century. For a western regional perspective on these issues, see Richard White, *"It's Your Misfortune and None of My Own": A New History of the American West* (Norman: University of Oklahoma Press, 1991). For works that focus on the implications of early industrialization on the formation of environmental politics and policy, see Richard N. L. Andrews, *Managing the Environment, Managing Ourselves: A History of American Environmental Policy* (New Haven: Yale University Press, 1999); John Opie, *Nature's Nation: An Environmental History of the United States* (Fort Worth: Harcourt Brace College Publishers, 1998), and Benjamin Kline, *First Along the River: A Brief History of the U.S. Environmental Movement*, 4th ed. (Lanham: Rowman & Littlefield, 2011).

4. Roderick Nash, *Wilderness and the American Mind*, 3rd ed. (New Haven: Yale University Press, 1982).

5. These estimates come from the U.S. Department of the Interior, U.S. Fish and Wildlife Service, "Time Line of the American Bison," http://www.fws.gov/bisonrange/timeline.htm.

6. White, *"It's Your Misfortune and None of My Own."*

7. Michael Dombeck, Christopher Wood, and Jack Williams, *From Conquest to Conservation: Our Public Lands Legacy* (Washington, DC: Island Press, 2003), 16.

8. See Stephen Pyne, *Fire: A Brief History* (Seattle: University of Washington Press, 2001).

9. George Catlin, *Manners, Customs and Conditions of the North American Indians*, in *American Earth: Environmental Writing Since Thoreau*, ed. Bill McKibben (New York: Library of America, 2008), 37–45. Originally published in 1841.

10. Robert Gottlieb, *Forcing the Spring: The Transformation of the American Environmental Movement* (Washington, DC: Island Press, 1993).

11. See Jonathan H. Adler, "Stand or Deliver: Citizen Suits, Standing, and Environmental Protection," *Duke Environmental Law and Policy Forum* 12, no. 39 (2001): 39–83.

12. This appeared in the June 5, 1990, issue of *Time.*

13. The 1994 California Desert Lands Act stands as one significant exception.

14. See William Chaloupka, "The County Supremacy and Militia Movements: Federalism as an Issue on the Radical Right," *Publius* 26, no. 3 (1996): 161–75.

15. Erik Larson/Tonopah, "Unrest in the West: Nevada's Nye County," *Time*, October 23, 1995, 52–60. For a more complete discussion of the regional transitions comprising the New West, see William Travis, *New Geographies of the American West: Land Use and Changing Patterns of Place* (Washington, DC: Island Press, 2007)

and James Robb et al., *Atlas of the New West: Portrait of a Changing Region* (New York: W. W. Norton, 1997). For a discussion of how these socioeconomic dynamics informed the emergence of collaborative conservation in the United States, see Randall K. Wilson, "Narratives of Community-Based Resource Management in the American West," in *Rural Change and Sustainability*, ed. Stephen Essex et al. (Wallingsford, UK: CABI Publishing, 2005), 343–57, and "Collaboration in Context: Rural Change and Community Forestry in the Four Corners," *Society and Natural Resources* 19 (2006): 53–70.

16. For a more detailed account of the rise of collaborative conservation, including in-depth treatments of a number of case studies, see Julia Wondollek and Steven Yaffee, *Making Collaboration Work: Lessons from Innovation in Natural Resource Management* (Washington, DC: Island Press, 2000) and Robert B. Keiter, *Keeping Faith with Nature: Ecosystems, Democracy, and America's Public Lands* (New Haven: Yale University Press, 2003). Collaborative conservation has also been attempted in other contexts, including efforts to resolve state-level and private land use conflicts involving historical Mexican and Spanish land grant claims. See, for example, Randall K. Wilson, "'Placing Nature': The Politics of Collaboration and Representation in the Struggle for La Sierra in San Luis, Colorado," *Ecumene* 6 (1999): 1–28.

17. Steven W. Selin, Michael A. Schuett, and Deborah S. Carr, "Has Collaborative Planning Taken Root on the National Forests?" *Journal of Forestry* 96 (1997): 25–28.

18. See Wondollek and Yaffee, *Making Collaboration Work*, as well as Mark Baker and Jonathan Kusel, *Community Forestry in the United States: Learning from the Past, Crafting the Future* (Washington, DC: Island Press, 2003).

19. Randall K. Wilson, "Community-Based Management ands National Forests in the Western United States: Five Challenges," *Policy Matters* 12 (2003): 216–24.

20. Suzanne Goldenberg, "The Worst of Times: Bush's Environmental Legacy Examined," *Guardian*, January 16, 2009.

CHAPTER 4: NATIONAL PARKS

1. John Muir, *The Yosemite* in *The Eight Wilderness Discovery Books* (Seattle: The Mountaineers, 1992), 714. First published in 1911.

2. This figure is constantly changing. However, the number presented here (401) represents the reported total number of units in the national park system as of April 2013, according to the National Park Service. See National Park Service, U.S. Department of the Interior, "National Park Service Overview," http://nps.gov/news/upload/NPS-Overview-2012_updated-04-02-2013.pdf.

3. For a detailed account of such events as they relate to the cases of Yellowstone, Yosemite, and Glacier National Park, see Mark D. Spence, *Dispossessing the Wilderness: Indian Removal and the Making of the National Parks* (New York: Oxford University Press, 1999).

4. A detailed account of Euro-American encounters with Yellowstone in the 1800s is found in Aubrey L. Haines, *The Yellowstone Story*, volumes 1 and 2, rev. ed. (Boulder: University Press of Colorado, 1996 and 1999).

5. As mentioned in chapter 3, as early as colonial times, government authorities prohibited private ownership of certain lands containing valuable resources,

such as timber stands or mineral deposits. In 1832, Congress set aside 640 acres to protect the Arkansas Hot Springs as a "reservation," but early residents generally ignored this classification. Similarly, the 1864 law protecting Yosemite Valley and Mariposa Grove of giant sequoias predates Yellowstone, as does President Grant's action in 1868 to set aside the Pribilof Islands in Alaska to protect the northern fur seal rookery. Though both Yosemite Valley and Arkansas Hot Springs later became national parks, these redesignations did not occur until 1906 and 1921, respectively. For more on the Pribilof Islands, see U.S. Department of Commerce, National Oceanic and Atmospheric Association, *Pribilof Islands: A Historical Perspective,* http://docs.lib.noaa.gov/noaa_documents/NOS/ORR/ TM_NOS_ORR/TM_NOS-ORR_17/HTML/Seal_Islands.htm.

6. See Alfred Runte, *National Parks: The American Experience,* 4th ed. (Lanham: Rowman & Littlefield, 2010): 43–55. Earlier expressions of this thesis are found in Alfred Runte, "Yellowstone: It's Useless, So Why Not a Park?" *National Parks and Conservation Magazine: The Environmental Journal* 46 (March 1972): 4–7 and "'Worthless' Lands: Our National Parks," *American West* 10 (May 1973): 4–11.

7. For a more detailed account of this idea, see Chris J. Magoc, *Yellowstone: The Creation and Selling of an American Landscape* (Albuquerque: University of New Mexico Press, 1999), and Mark Daniel Barringer, *Selling Yellowstone: Capitalism and the Construction of Nature* (Lawrence: University Press of Kansas, 2002).

8. Richard White, *"It's Your Misfortune and None of My Own": A New History of the American West* (Norman: University of Oklahoma Press, 1991).

9. Congress authorized the restoration and protection of the Casa Grande Ruins in Arizona in 1889 in order to preserve prehistoric structures dating from the thirteenth century. President Benjamin Harrison later set aside the ruins as a reservation in 1892, but they did not gain national monument status until 1918. They were placed on the National Register of Historic Places in 1966.

10. In addition to Muir's own accounts of his life and understandings of nature, an excellent biography that traces the evolution of Muir's thinking is provided in Donald Worster, *A Passion for Nature: The Life of John Muir* (New York: Oxford University Press, 2008).

11. Dayton Duncan and Ken Burns, *The National Parks: America's Best Idea* (New York: Knopf Doubleday, 2009).

12. There is much more to the story of Mesa Verde National Park. Recall that originally the land was located on the Ute Mountain Ute Reservation. Given the thinking of the day, national park status required the removal of indigenous peoples and the relinquishment of any land claims to conform to the notions of "empty and eternal," not to mention the claim of federal sovereignty. And all of this must occur despite the fact that the primary rationale for federal protection was premised on artifacts left by ancient indigenous peoples. In the end, the federal government successfully negotiated for Mesa Verde by swapping lands encompassing Sleeping Ute Mountain, an area deemed sacred to the Ute people. See Duane A. Smith, *Mesa Verde National Park* (Charleston, SC: Arcadia Publishing, 2009).

13. The six national monuments declared by Teddy Roosevelt that later transformed into national parks include the following (with national park authorization dates listed in parentheses): Lassen Peak and Cinder Cone, later redesignated as Lassen Volcanic National Park (1916), Grand Canyon National Park (1919),

Olympic National Park (1938), Petrified Forest National Park (1962), Chaco Canyon National Historic Park (1980), and Kino Missions, later renamed Tucumcari National Historic Park (1990).

14. John Muir, *The Yosemite*, 714.

15. Duncan and Burns, *The National Parks*.

16. Paul Sutter, *Driven Wild: How the Fight Against the Automobile Launched the Modern Wilderness Movement* (Seattle: University of Washington Press, 2002).

17. The data presented here derive from the National Park Service, "Top 10 Visited Sites in National Park System Revealed (2012)," Press release issued April 3, 2013. Available at www.nps.gov/news/release.htm?id=1457.

18. Unit types refers to all the different types of protected areas in the national park system, including monuments, lakeshores, recreation areas, historic trails, and more. The Blue Ridge Parkway is a national parkway managed by the National Park Service.

19. National Park Service, "Top 10 Visited Sites in National Park System Revealed (2012)."

20. These five included Yellowstone, Rocky Mountain, Great Smoky, Shenandoah, and Yosemite National Parks.

21. However, victory was tempered by a compromise deal that led to the expansion of the Glen Canyon Dam in 1956. Brower vowed to never allow such a compromise in future campaigns. For a thoughtful and detailed account of this conflict, see Mark Harvey, *A Symbol of Wilderness: Echo Park and the American Conservation Movement* (Albuquerque: University of New Mexico Press, 1994).

22. More specifically, the lands from which the state of Alaska could choose were those managed by the Bureau of Land Management.

23. This declaration took place in October 1979.

24. President Bush's national marine reserves and national marine monument declarations covered a larger area in terms of acreage, but they are comprised almost entirely of "submerged lands" on the ocean floor.

25. Since 1906, only three presidents refused to use the Antiquities Act to create new national monuments: presidents Nixon, Reagan, and George H. W. Bush. See the Wilderness Society, *The Antiquities Act: Protecting America's Natural and Cultural Treasures*, March 14, 2011, http://wilderness.org/resource/antiquities-act-protecting-americas-natural-cultural-treasures.

26. It is interesting to note that the national park status of American Samoa National Park is premised on a fifty-year lease from the local government paid by the U.S. federal government amounting to $537,000 per year (a little over one-fourth of the annual operating costs). The governor of American Samoa then makes payments to villagers who still retain traditional title to the lands inside the park boundaries. The park is 13,500 acres in size, four thousand acres of which are in marine environments.

27. See William C. Everhart, *The National Park Service* (Boulder: Westview Press, 1983), 107–21.

28. As part of a broader set of Readiness and Range Initiatives, the administration argued at the time that these regulatory changes were needed for national security reasons. The military needed to be released from compliance with various environmental laws in order to properly prepare for national defense. In contrast,

the marine national monuments do not interfere with military preparations. Environmental organizations were skeptical of this reasoning at the time and remain skeptical to this day, arguing that military readiness can be achieved without undermining national environmental laws.

29. Recall that Charles Young was the first black man to serve as a superintendent in the national park system. Both the Rio Grande del Norte and San Juan Islands monuments were carved out of BLM lands, and that agency will continue to manage them. The site in Delaware is the first-ever national monument in that state. With this declaration, all fifty states now contain national monuments. The site celebrates the colonial and indigenous heritage of the region. For more on these declarations, see Sasha Ingber, "Obama Declares Monuments to Preserve Pieces of U.S. Heritage," *National Geographic News*, March 23, 2013, http://news .nationalgeographic.com/news/2013/02/130326-national-monuments-obama -nation-science-travel/.

30. See Karen E. Bagne and Deborah M. Finch, "Vulnerability of Species to Climate Change in the Southwest: Threatened, Endangered, and At-Risk Species at the Barry M. Goldwater Range, Arizona," *General Technical Report RMRS-GTR-284* (Fort Collins, CO: U.S. Department of Agriculture, Forest Service, Rocky Mountain Research Station, 2012): 1–139.

31. M. H. P. Hall and D. B. Fagre, "Modeled Climate-Induced Glacier Change in Glacier National Park, 1850–2100," *Bioscience* 53 (2003): 131–40. See also the information gathered by the U.S. Geological Survey on this topic at http://www. nrmsc.usgs.gov/research/glacier_retreat.htm.

32. National Park Service, "Top 10 Visited Sites in National Park System Revealed (2012)."

33. For a detailed account of the many types of issues facing park managers, see Robert E. Manning and Laura E. Anderson, *Managing Outdoor Recreation: Case Studies in the National Parks* (Cambridge, MA: CABI, 2012).

34. The Fund for Animals and Biodiversity Legal Foundation filed the suit.

35. These activities refer to annual hunts, reminiscent of wildlife management practices on national refuges or other forms of slaughter. Until the 1960s, the animals were managed as domesticated livestock rather than wildlife.

36. Nate Schweber, "2 Bills Propose Zero Tolerance for Bison," *New York Times*, February 1, 2013, http://green.blogs.nytimes.com/2013/02/01/2-bills-propose -zero-tolerance-for-bison/.

37. Kirk Johnson, "Deal Puts Yellowstone Bison on Ted Turner's Range," *New York Times*, May 21, 2010, http://www.nytimes.com/2010/05/22/us/22bison .html?_r=0.

38. Ibid.

CHAPTER 5: NATIONAL FORESTS

1. The quote actually comes from a letter signed by Department of Agriculture secretary James Wilson, dated February 1, 1905. But most national forest historians believe that Pinchot was the author.

2. As of June 2012, the Forest Service administers over thirty-six million acres of land designated as Wilderness Areas. The National Park Service manages nearly forty-four million acres, the Fish and Wildlife Service over twenty million, and the BLM 8.7 million acres.

3. This is approximately 74 percent of the 276 million visitors to the national parks that same year.

4. This statement is accurate in the context of the federal agencies examined in this book. There are a few other exceptions to the Department of the Interior rule. For example, the Army Corps of Engineers is located in the U.S. Department of Defense and plays a significant role in the construction and management of dam projects that directly impact both public and private lands in the United States. But given the agency's emphasis on regulating waterways as opposed to federal lands per se, neither the Army Corps (nor the related Bureau of Reclamation) is examined here.

5. George Perkins Marsh, *Man and Nature, or Physical Geography as Modified by Human Action* (New York: Charles Scribner & Co., 1867), 36.

6. Harold Steen, *The U.S. Forest Service: A History, Centennial Edition* (Seattle: Forest History Society and University of Washington Press, 2004), 9–20.

7. Michael Dombeck, Christopher Wood, and Jack Williams, *From Conquest to Conservation: Our Public Lands Legacy* (Washington, DC: Island Press, 2003), 16. See also Reverend Peter Pernin, *The Great Peshtigo Fire: An Eyewitness Account*, 2nd ed. (Madison: Wisconsin Historical Society, 1999).

8. Steen, *The U.S. Forest Service*, 26–28.

9. James G. Lewis, *The Forest Service and the Greatest Good: A Centennial History* (Durham: Forest History Society, 2006), 18–22.

10. Steen, *The U.S. Forest Service*, 30–34.

11. Dyan Zaslowsky and T. H. Watkins, *These American Lands: Parks, Wilderness, and the Public Lands* (Washington, DC: Wilderness Society and Island Press, 1994), 74.

12. See Lewis, *The Forest Service and the Greatest Good*, 36–42.

13. Gifford Pinchot, *The Fight for Conservation* (New York: Doubleday, Page, 1910), 43–46.

14. See note 1. However, the idea is paraphrased in Pinchot's *The Fight for Conservation*, 1910.

15. For the classic text on the development and implementation of the idea of progressive conservation, see Samuel P. Hays, *Conservation and the Gospel of Efficiency: The Progressive Conservation Movement* (Albuquerque: University of New Mexico Press, 1959).

16. In addition, this law formally changed the name of the national forest reserves to national forests.

17. Though it was only the latest in a long string of challenges to President Taft's decision making with regard to the public domain, Pinchot's public criticism of the Secretary of the Interior was the straw that broke the camel's back. For more on how this all played out, see Steen, *The U.S. Forest Service*, 100–4, and Lewis, *The Forest Service and the Greatest Good*, 64–68.

18. A more detailed and extremely readable account of these events, including the impact of the fire on national forest policy, is found in Timothy Egan, *The Big*

Burn: Teddy Roosevelt and the Fire That Saved America (New York: Houghton Mifflin Harcourt, 2009).

19. Lewis, *The Forest Service and the Greatest Good*, 79.

20. Ibid.

21. Zaslowsky and Watkins, *These American Lands*, 82–83. The authors also point to the 1952 assistant chief forester Christopher Granger as one who was pleased about the prospect of the Forest Service operating "in the black," as though it was a business venture. Granger authored an article in *American Forests* to this effect, underscoring the benefits of increased production.

22. Ibid.

23. Dombeck et al., *From Conquest to Conservation*, 36.

24. Muir may have lost the battle for Hetch Hetchy, but he won the battle of public opinion. Several authors note that the public support for the Forest Service appeared to dip following the debate. See Lewis, *The Forest Service and the Greatest Good*, 111–2, and Steen, *The U.S. Forest Service*, 114–5.

25. Some examples include the Grand Canyon National Park in Arizona in 1919, Olympic National Park in Washington in 1938, and Kings Canyon National Park in California in 1940. It is also worth noting the differences in approach between national parks and forest vis-à-vis expansion in the eastern United States. While the national forests were founded on lands purchased from willing sellers, new national parks often relied on governmental claims of eminent domain, often by state governments, to acquire eastern territories. For an excellent account in the Appalachians, see Kathryn Newfont, *Blue Ridge Commons: Environmental Activism and Forest History in Western North Carolina* (Athens: University of Georgia Press, 2012).

26. Success of the national campaign was evidenced with the owl's appearance on the front cover of *Time* on June 5, 1990. However, it is worth noting that both sides sought broader national-scale support.

27. See Judith A. Layzer, *The Environmental Case: Translating Values into Policy*, 3rd ed. (Washington, DC: CQ Press, 2012), 182–83. Much of the discussion here regarding legalistic details of the northern spotted owl controversy derives from Layzer's account.

28. The Timber Salvage Rider allowed logging in areas protected as critical owl habitat but which were damaged by fire, disease, or insect outbreaks. Environmentalists feared the rider would be abused to undermine ESA protections, opening the door to continued logging under questionable circumstances. The 2003 Healthy Forest Act was similar, allowing logging in areas previously protected from timber harvests in the name of reducing fire, insect, and disease outbreaks on national forests.

29. For detailed accounts of the transformation in national forest management toward a more conservation-based, ecosystem management approach in the 1980s and 1990s, see Paul Hirt, *A Conspiracy of Optimism: Management of the National Forests Since World War Two* (Omaha: University of Nebraska Press, 1994) and Samuel P. Hays, *Wars in the Woods: The Rise of Ecological Forestry in America* (Pittsburgh: University of Pittsburgh Press, 2009). For more on the emergence of collaborative conservation on national forests, see Julia Wondolleck and Steven Yaffee, *Making Collaboration Work* (Washington, DC: Island Press, 2000)

and Jonathan Kusel and Elisa Adler, eds., *Forest Communities, Community Forests* (Lanham: Rowman & Littlefield, 2003).

30. See Gerald W. Williams, *The Forest Service: Fighting for Public Lands* (Westport: Greenwood Press, 2007), 202.

31. Bark beetle outbreaks have impacted hundreds of thousands of acres of pine forests in the American West in recent decades. While ecologists believe the primary cause is linked to drought conditions and climate change impacts that have weakened the resistance of various tree species while creating ideal conditions for spikes in beetle populations, even-age management is also cited as exacerbating the effects of these outbreaks. See U.S. Forest Service, "Western U.S. Bark Beetles and Climate Change," http://www.fs.fed.us/ccrc/topics/insect -disturbance/bark-beetles.shtml.

32. Hays, *Wars in the Woods*, 2009. See also Samuel Hays, *The American People and the National Forests: The First Century of the U.S. Forest Service* (Pittsburgh: University of Pittsburgh Press, 2009): 106–36. Interestingly, Hays does not address the role of nontimber forest products (NTFPs) as another growing area of forest use. See Marla Emery and Rebecca McLain, *Non-Timber Forest Products: Medicinal Herbs, Fungi, Edible Fruits and Nuts, and Other Natural Products from the Forest* (Binghamton: Food Products Press, 2001).

33. See Ed Marston, "The Quincy Library Group: A Divisive Attempt at Peace," in *Across the Great Divide: Explorations in Collaborative Conservation and the American West*, ed. Philip Brick et al. (Washington, DC: Island Press, 2001), 79–90.

34. Sierra Pacific Industries owned and operated the largest mills in the area.

35. The County Supremacy Movement shared much ideologically with the Sagebrush Rebellions of the 1940s and 1970s and the Wise Use Movement of the 1990s. However, rather than involve the states, County Supremacists simply declared all federal lands within their borders as "county property." However, none of these claims held up in court. See William Chaloupka, "The County Supremacy and Militia Movements: Federalism as an Issue on the Radical Right," *Publius* 26, no. 3 (1996): 161–75.

36. Another 30 percent of the county land area is comprised of the Ute Mountain Ute Indian Reservation and state-owned lands. Private property accounts for approximately 28 percent of Montezuma County.

37. For more on the differences between the PPFP, the Greater Flagstaff Forest Partnership (GFFP), and the Catron County Citizens Group (CCCG), see Randall K. Wilson, "Collaboration in Context: Rural Change and Community Forestry in the Four Corners," *Society and Natural Resources* 19 (2006): 53–70. A detailed account of the CCCG can be found in Sam Burns, "Catron County, New Mexico: Mirroring the West, Healing the Land, Rebuilding Community," in *Forest Communities, Community Forests*, ed. Jonathan Kusel and Elisa Adler (Lanham: Rowman & Littlefield, 2003): 89–115.

38. Dombeck et al., *From Conquest to Conservation*, 103.

39. Ibid. But also see U.S. Department of Agriculture, Forest Service, "FY 1905–2012 National Summary Cut and Sold Data," http://www.fs.fed.us/forest-management/products/sold-harvest/index.shtml.

40. In the 1950s, the Forest Service signed unique fifty-year contracts with two large logging companies, guaranteeing a certain amount of timber each year in

exchange for investment in two large pulp mills in Ketchikan and Sitka. The only contracts of their kind in the United States, they prioritized timber production over all other management priorities on the Tongass. The 1990 Tongass Reform Act limited the amount of annual timber harvests and offered other conservation reforms, but it left the contracts in place. In 1993 and 1997, respectively, the two contracts were finally terminated. See Douglas Chadwick, "The Truth about Tongass," *National Geographic*, July 2007, http://ngm.nationalgeographic.com/ngm/0707/feature4/index.html.

CHAPTER 6: NATIONAL WILDLIFE REFUGES

1. Aldo Leopold, *A Sand County Almanac, with Other Essays on Conservation from Round River* (New York: Oxford University Press and Ballantine Books, 1966), 138. First published in 1949.

2. This number is constantly changing. As of September 30, 2012, there were 560 units in the refuge system. These include wildlife refuges, wildlife ranges, wildlife management areas, game preserves, and conservation areas. The data are drawn from Department of the Interior, U.S. Fish and Wildlife Service, *Annual Report of Lands under Control of the U.S. Fish and Wildlife Service* (Washington, DC: GPO, 2012).

3. Attempting to identify and quantify the lands administered by the U.S. Fish and Wildlife Service is an extremely complicated task. The huge variation in the amount of national monuments all depends if one counts only the National Marine Refuges for which the USFWS is the sole manager (approximately fifty million acres), or all of the units in the South Pacific that it comanages with other federal agencies or administrative bodies (over 150 million acres).

4. However, it is important to point out that even when certain species may have provided the initial impetus for creating a national wildlife refuge (as in the two examples listed here), today the mandate for the USFWS includes the protection and enhancement of biodiversity writ large. Hence, officials manage for a huge variety of species native to local ecosystems.

5. Department of the Interior, U.S. Fish and Wildlife Service, *America's National Wildlife Refuge System: Hunting and Fishing*, Pub. 9 (Washington, DC: GPO, 2003).

6. All of these figures derive from Department of the Interior, U.S. Fish and Wildlife Service, *Annual Report of Lands under Control of the U.S. Fish and Wildlife Service*. National Wildlife Refuges Coordination Areas are managed by state agencies under cooperative agreements with the USFWS. Administrative sites are maintenance facilities, offices, and visitor centers that are owned by the USFWS but located outside of refuge boundaries.

7. The other federal agency mandated to enforce the ESA is the National Marine Fisheries Service, part of the National Oceanic and Atmospheric Administration (NOAA) located in the Department of Commerce. As the name implies, this agency is responsible for all marine species under the act.

8. It is important to note that in cases where private property is involved, a Habitat Protection Plan (HCP) is crafted in conjunction with the landowner to outline the best way forward. Such plans can include modifications to the

habitat and even the "taking" of protected species, as long as mediation arrangements are put into place.

9. Department of the Interior, U.S. Fish and Wildlife Service, *National Wildlife Refuge System Fact Sheet* (Washington, DC: GPO, 2011).

10. On this point, see Jeanne Nienaber Clarke and Daniel C. McCool, *Staking Out the Terrain: Power Differentials among Natural Resource Management Agencies*, 3rd ed. (Albany: State University of New York Press, 1996). Their focus is on the U.S. Fish and Wildlife Service rather than the national wildlife refuge system per se, but their observations apply to both the lands and the agency.

11. This statement is not meant to imply that it was "easy" to establish Grand Canyon or Yellowstone National Parks. Clearly, these efforts involved significant political struggles as outlined in chapter 4. However, unlike the national park debates, which are often, though not always, largely settled with congressional legislation, the political battles over the fate of wildlife (such as the northern spotted owl or grey wolf) seem to defy a clear and final resolution. The difference may be a function of the diverse historical and political eras within which these conflicts and struggles took place. However, perhaps it also has something to do with the different ways we tend to conceptualize, value, and define landscapes as opposed to wildlife?

12. Michael Bean and Melanie Rowland, *The Evolution of National Wildlife Law*, 3rd ed. (Westport: Praeger, 1997).

13. Paul Wilson, "The Public Trust Doctrine in Wildlife," *Mountain State Sierran Newsletter*, Sierra Club, 2005.

14. Quoted in Dyan Zaslowsky and T. H. Watkins, *These American Lands: Parks, Wilderness, and the Public Lands* (Washington, DC: Wilderness Society and Island Press, 1994), 159.

15. Ibid., 162.

16. As mentioned earlier, it is true that President Grant established the Pribilof Islands reserve in Alaska in 1868 in order to protect the northern fur seal rookery from international hunters, but it did not prohibit access to American seal hunting companies.

17. Historians point to several earlier prototype refuges. The 1864 transfer of Yosemite Valley to the state of California included a clause forbidding the "wanton destruction" of wildlife in the park. A similar clause was contained in the legislation creating Yellowstone National Park in 1872, though as we know, this did little to curb such practices. In 1869, Congress passed a law protecting Alaska's Pribilof Islands as a reserve for northern fur seals. And in 1892, President Harrison declared Afognak Island in Alaska to be a "fish, cultural and forest reserve" under the Forest Reserves Act. However, these actions were designed primarily to protect the commercially valuable seal population and fisheries resources for the American fishing industry against inroads by other countries. Hence, one can argue that in contrast to Roosevelt's action at Pelican Island, the primary motivation for the Alaskan reserves was not tied to conservationist goals, but rather driven by commercial interests.

18. Quoted in Zaslowsky and Watkins, *These American Lands*, 167.

19. The Duck Stamp Act is formally known as the Migratory Bird Hunting and Conservation Stamp Act of 1934.

20. For a more detailed account, see George Laycock, *The Sign of the Flying Goose: A Guide to the National Wildlife Refuges* (Garden City: Doubleday, 1965).

21. National Wildlife Federation, *Creation of the National Wildlife Federation*, 2012, www.nwf.org/who-we-are/. A year later, in 1937, another influential conservation organization came into being, calling itself Ducks Unlimited. But unlike the NWF, this group focused more specifically on issues related to waterfowl habitat protection and sports hunting.

22. Clarke and McCool, *Staking Out the Terrain*, 107–9.

23. Department of the Interior, U.S. Fish and Wildlife Service, "Short History of the National Wildlife Refuge System" (Washington, DC: GPO, 2012), www.fws .gov.history.

24. For a detailed history of the USFWS predator control program, see Donald Hawthorne, "The History of Federal and Cooperative Animal Damage Control," *Sheep and Goat Research Journal* Paper 6, University of Nebraska-Lincoln, 2004.

25. Ibid., 13–14.

26. Perhaps nothing reflects the tensions of this legacy more than the agency's role in restoring wolf populations in the northern Rockies in the 1990s. A species the USFWS once worked to eradicate now receives agency protection under the Endangered Species Act.

27. Clarke and McCool, *Staking Out the Terrain*, 114.

28. Zaslowsky and Watkins, *These American Lands*, 174–82.

29. Linda Lear, *Rachel Carson: Witness for Nature* (New York: Henry Holt, 1997). See also Department of the Interior, U.S. Fish and Wildlife Service, "History of the National Wildlife Refuge System" (Washington, DC: GPO, 2012), www.fws .gov/history.

30. Ibid.

31. National Wildlife Federation, "List of National Wildlife Refuges Created by Presidential and Congressional Action," 2004, www.nwf.org.

32. Two of President Obama's wildlife refuges include conservation areas as well. A wildlife refuge conservation area (distinct from a BLM-administered conservation area) is akin to a conservation easement. This is land in which the USFWS may hold an easement or some other partial interest to prevent development while the existing property owner continues to hold title.

33. Zaslowsky and Watkins, *These American Lands*, 182.

34. Department of the Interior, U.S. Fish and Wildlife Service, "Legislative Mandates and Authorities for the U.S. Fish and Wildlife Service," 2012, www .fws.gov DOI.

35. Department of the Interior, U.S. Fish and Wildlife Service, "Everglades Headwater National Wildlife Refuge and Conservation Area," 2012, www.fws .gov/southeast/evergladesheadwaters.

36. Department of the Interior, U.S. Fish and Wildlife Service, "The Arctic National Wildlife Refuge."

37. Rachel D'Oro, "Alaska Residents Cash In on Annual Dividend," *Seattle Sun Times*, September 5, 2008.

38. This scenario is playing out with the recent boom in Marcellus Shale natural gas development. See John Roberts, "Plan for US Natural Gas Exports Brings Talk of Economic Boon, Fear of Failure," *Fox News*, December 17, 2012.

39. Phillip Shabecoff, "US Proposing Drilling for Oil in ANWR," *New York Times*, November 25, 1986. Quoted in Judith A. Layzer, *The Environmental Case: Translating Values into Policy*, 3rd ed. (Washington, DC: CQ Press, 2012), 121.

40. Layzer, *The Environmental Case*, 125.

41. Ibid.

42. Department of the Interior, U.S. Fish and Wildlife Service, *The National Elk Refuge: A Legacy of Conservation* (Washington, DC: GPO, 2008).

43. Ibid.

44. Jim Robbins, "Suit Opposes Elk Feeding in Wyoming," *New York Times*, June 4, 2008.

45. Ibid.

46. Timothy Preso, "Court Rules Continuing Supplemental Feeding on the National Elk Refuge Undermines Conservation," *Earthjustice Press Release*, August 4, 2011, http://earthjustice.org/news/press/2011/court-rules-continuing-supplemental-feeding-on-the-national-elk-refuge-undermines-conservation.

47. Department of the Interior, U.S. Fish and Wildlife Service, *A Description of the National Elk Refuge and Grand Teton National Park* (Washington, DC: GPO, 2007).

CHAPTER 7: BUREAU OF LAND MANAGEMENT LANDS

1. See Samuel T. Dana and Sally K. Fairfax, *Forest and Range Policy* (New York: McGraw-Hill, 1980).

2. U.S. Department of the Interior, Bureau of Land Management, *Public Land Statistics 2011* (2012). The subsurface deposits administered by the BLM lie under all types of federal lands, including state lands and, in some cases, even privately owned property.

3. Ibid.

4. I refer here to the 2.6-million-acre Oregon and California Railroad Revested Lands (O & C lands), which fell to the BLM after railroad companies charged with land fraud returned them to the federal government in 1916. The 1937 O & C Act called for sustainable timber production on these lands, with a requirement that local counties receive 50 percent of timber receipts in lieu of lost property taxes.

5. See Jeanne N. Clarke and Daniel McCool, *Staking Out the Terrain: Power Differentials among Natural Resource Management Agencies*, 2nd ed. (Albany: State University of New York Press, 1996), 157–8. The point is also made by Dana and Fairfax, *Forest and Range Policy*, 229.

6. Benjamin H. Hibbard, *A History of the Public Land Policies* (New York: Macmillan, 1965).

7. Richard N. L. Andrews, *Managing the Environment, Managing Ourselves: A History of American Environmental Policy* (New Haven: Yale University Press, 1999), 103–4.

8. White, *"It's Your Misfortune and None of My Own": A New History of the American West* (Norman: University of Oklahoma Press, 1991), 221–5.

9. Author Frances Keck notes that the JJ Ranch in Colorado controlled approximately 960,000 acres by holding title to only eighteen thousand acres through various homestead acts. Article available at www.coloradopreservation.org.

10. Western historian Richard White (*"It's Your Misfortune and None of My Own,"* 224) suggests the 30 percent figure, Dyan Zaslowsky and T. H. Watkins (*These American Lands: Parks, Wilderness, and the Public Lands* (Washington, DC: Wilderness Society and Island Press, 1994), 117) posit 75 percent, while Charles Wilkinson (*Crossing the Next Meridian: Land, Water and the Future of the West* (Washington, DC: Island Press, 1992)) estimates 90 percent of domestic herds across the entire public domain.

11. These refer to the 1890 creation of three national parks in California (General Grant, Sequoia, and Yosemite), the 1894 Yellowstone Game Protection Act, and the 1894 Carey Act.

12. Initially designed to protect fuel supplies for the U.S. Navy, this forward-thinking law established a regulatory framework for the mining of coal, oil, natural gas, oil shale, and other fossil fuels as well as fertilizers (phosphates, potassium, and sodium). Recognizing their value, the development of these materials on public lands now required leases and royalty payments to the federal government (unlike the hard rock minerals covered under the 1872 General Mining Law).

13. The scandal involved Albert Fall, Interior Secretary in the Harding administration. Fall was convicted of taking bribes from oil companies in return for illegally securing their access to oil reserves in the public domain, including the Teapot Dome formation in Wyoming and Elk Hills in California. Although the actions took place in 1922, the conviction occurred in 1929. See Zaslowsky and Watkins, *These American Lands*, 120–1.

14. See Donald Worster, *The Dust Bowl: The Southern Plains in the 1930s* (New York: Oxford University Press, 1979). The worst of these events occurred on April 14, 1935. Known as Black Sunday, over twenty storms coalesced across the plains on this day, darkening the skies in the afternoon.

15. The measure is based on a cow weighting one thousand pounds and amounts to approximately eight hundred pounds of dry forage. For a mature sheep, the amount is approximately 20 percent of the amount determined for cattle. These measures help producers and managers discern the carrying capacity of the land and serve as a standard for setting grazing fees. See M. Pratt and G. Rasmussen, "Determining Your Stocking Rate," *Utah State University Cooperative Extension* NR/RM/04, 2001.

16. Wilkinson, *Crossing the Next Meridian*.

17. Judith A. Layzer, *The Environmental Case: Translating Values into Policy*, 3rd ed. (Washington, DC: CQ Press, 2012), 143. The idea here is that a permit system guaranteed access to the lands by ranchers and protected them against intrusions by homesteaders or competing livestock producers.

18. The Soil Conservation Service was the forerunner of today's Natural Resource Conservation Service (NRCS).

19. Text from the 1934 Taylor Grazing Act, 43 USC Section 315, Title 43, Chapter 8A, Subchapter 1.

20. This was confirmed via executive order by FDR in 1935.

21. Generally speaking, the Homestead Acts remained in force in Alaska until 1976. Moreover, the General Mining Law still allows mining claims on the public domain to this day.

22. On the one hand, this was a seemingly odd choice given that the historical role of the General Lands Office (GLO) was to oversee the transfer of the public domain into private hands. Corruption in the GLO through the nineteenth century did much to facilitate many of the worst abuses of public land laws in terms of speculation and environmental degradation. However, since the 1870s the GLO was also involved in classifying lands for their so-called best use under the various Homestead Acts and federal grant programs, and as such it was engaged in the process of identifying lands for inclusion in the grazing districts mandated by the Taylor Act.

23. See Zaslowsky and Watkins, *These American Lands*, 124–5. See also Marion Clawson, *The Federal Lands Revisited* (Washington, DC: Resources for the Future, 1983), 36–7. Clawson points out that in addition to being a rancher, Carpenter was also a Harvard-trained lawyer.

24. For a full discussion of the shift in Ickes's views on this matter, see Karen Merrill, *Public Lands and Political Meaning: Ranchers, the Government and the Property between Them* (Berkeley: University of California Press, 2002).

25. Paul J. Culhane, *Public Land Politics: Interest Group Influence on the Forest Service and Bureau of Land Management* (Baltimore: Johns Hopkins University Press, 1981). Quoted in Clarke and McCool, *Staking Out the Terrain*, 161.

26. Bernard DeVoto, "The West Against Itself," In *Western Paradox: A Conservation Reader*, ed. Douglas Brinkley and Patricia Nelson Limerick (New Haven: Yale University Press, 2000), 50–60.

27. James Skillen, *The Nation's Largest Landlord: The Bureau of Land Management in the American West* (Lawrence: University Press of Kansas, 2009), 23–24 and 30–34.

28. Dana and Fairfax, *Forest and Range Policy*, 229.

29. Ibid.

30. Interestingly, this authority was only temporary, lasting until the Public Land Law Review Commission issued its comprehensive report for public lands management (which occurred in 1970). Nonetheless, the BLM completed most of its classifications by 1969.

31. Skillen, *The Nation's Largest Landlord*, 48.

32. This 1974 lawsuit is not to be confused with the 1972 *NRDC v. Morton* case. The 1972 ruling ordered the BLM to offer credible alternatives when crafting an EIS (which, according to the court, it did not do when crafting an EIS for oil development leases off the coast of Louisiana). See Joan Ann Lukey, "NEPA's Impact Statement in the Federal Courts: A Case Study of *NRDC v. Morton*," *Boston College Environmental Affairs Law Review* 2 (1973): 807–26. In contrast, the 1974 suit dealt with the number of EIS's required by the BLM with regard to public land grazing. According to D. Dean Bibles, this decision also led to the BLM hiring a greater number of ecologists and resource management specialists. See D. Dean Bibles, "The Transforming Effect of the Natural Resources Defense Council Consent Decree," http://www.blm.gov/wo/st/en/info/history/sidebars/natural_resources/the_transforming_effect.print.html.

33. A detailed account of the various implications of FLPMA is found in Skillen, *The Nation's Largest Landlord*, 102–11.

34. Federal Land Policy and Management Act of 1976, 43 U.S.C. 1701, 1702, 1712.

35. Ibid.

36. A detailed account of these events is found in R. McGreggor Cawley, *Federal Land, Western Anger: The Sagebrush Rebellion and Environmental Politics* (Lawrence: University Press of Kansas, 1993).

37. Just as the Sagebrush Rebellion claimed state authority over the public domain, the County Supremacy movement claimed local (e.g., county) government authority over public lands. See William Chaloupka, "The County Supremacy and Militia Movements: Federalism as an Issue on the Radical Right," *Publius* 26, no. 3 (1996): 161–75.

38. These are the Oregon and California Revested Lands (O & C lands) mentioned earlier. The BLM was forced to dramatically curtail timber sales in the region in order to abide by the Northern Spotted Owl Recovery Plan as mandated under the Endangered Species Act. For this reason, the Northwest Forest Plan included federal payments to local counties to help replace income from lost timber receipts.

39. Recall that mining patents constitute the transfer of public mining claims to private ownership if certain conditions are met regarding annual investments and "improvements" according to the 1872 General Mining Law. This was always the Achilles' heel of the federal land system, a loophole that continuously allowed the fragmentation and loss of protected public lands and upset conservation management programs.

40. Kathy Rohling, ed., *Mining Claims and Sites on Federal Lands*, P-048, BLM National Science and Technology Center, revised May 2011.

41. The idea of creating new water rights on public lands can be confusing. In the past, ranchers who built artificial ponds for watering their livestock by damming or diverting rivers and streams on the public domain could then claim water rights to these "improvements." These rights guaranteed future access and, in some instances, established "sellable" assets that might increase the value of grazing permits and home ranches. As noted earlier, when ranches are sold, they often include grazing permits as part of the sale price. However, for the Clinton administration, these types of actions amounted to a subsidy to those permitted to graze livestock on public lands.

42. U.S. Department of the Interior, Bureau of Land Management, "Resource Advisory Councils." Information fact sheet last modified in 2011 and available at www.blm.gov. For more detail on the emergence of RACs as forums for BLM management, see William Riebsame, "Ending the Range Wars?" *Environment* 38, no. 4 (1996): 4–9, 27–29.

43. BLM Regions are organized along state boundaries. Some states have multiple regions (Colorado has three, for example), while other states, such as Arizona and the Dakotas, are served by a single region.

44. This was later expanded to 1.9 million acres.

45. Senator Orrin Hatch is quoted in Paul Larmer, "The Mother of All Land Grabs," *High Country News*, September 30, 1996, http://www.hcn.org/issues/90/2796.

46. Statement of Governor Michael Levitt, *Hearing before the House Subcommittee on National Parks and Public Lands: Establishing the Grand Staircase-Escalante National Monument*, 8(54), 1997. Quoted in Raymond B. Wrabely, "Managing the Monument: Cows and Conservation in Grand Staircase-Escalante National Monument," *Journal of Land, Resources and Environmental Law* 29, no. 2 (2009): 253–80.

47. Paul Larmer, "A Bold Stroke: Clinton Takes a 1.7 Million Acre Stand in Utah," *High Country News*, September 30, 1996, http://www.hcn.org/issues/90/2795.

48. U.S. Department of the Interior, Bureau of Land Management, *National Landscape Conservation Lands: Landscapes of the American Spirit* (Washington, DC: GPO, 2012).

49. Bruce Babbitt, "The Heart of the West: BLM's National Landscape Conservation System," in *From Conquest to Conservation: Our Public Lands Legacy*, by Michael P. Dombeck, Christopher A. Wood, and Jack E. Williams (Washington, DC: Island Press, 2003), 100–2.

50. For a detailed account, see Skillen, *The Nation's Largest Landlord*, 172–87. See also James Morton Turner, *The Promise of Wilderness: American Environmental Politics Since 1964* (Seattle: University of Washington Press, 2012), 367–71.

51. F. Barringer, "US to Open Public Land Near Parks for Drilling," *New York Times*, November 7, 2008. This action triggered lawsuits from an array of environmental groups, including the Sierra Club, Wilderness Society, Natural Resources Defense Council, Southern Utah Wilderness Alliance, and the National Park Conservation Association. See NPCA, "Protecting Arches and Canyonlands Parks," www.npca.org.

52. However, significantly, this did not include new appropriations for staffing or new recreation/conservation management programs.

53. J. Rudolf, "Federal Lands in Wyoming Opened to Coal Mining," *New York Times*, March 23, 2011.

54. J. Broder, "Interior Secretary Signs Grand Canyon Mining Ban," *New York Times*, January 9, 2012.

55. U.S. Department of the Interior, Bureau of Land Management, *Fact Sheet on the BLM's Management of Livestock Grazing*, http://www.blm.gov/wo/st/en/prog/grazing.html.

56. U.S. General Accounting Office, *Public Rangelands: Some Riparian Areas Restored but Widespread Improvements Will Be Slow* (Washington, DC: GAO, 1988).

57. See William Kittredge, "Home on the Range," *New Republic*, December 13, 1993, 13–16. Quoted in Layzer, *The Environmental Case*, 149. And of course, for some groups, including indigenous peoples, these classic cowboy images may represent something else entirely, standing as symbols of conquest and oppression. For the classic text on diversifying western history, see Patricia Nelson Limerick, *Legacy of Conquest: The Unbroken Past of the American West* (New York: W. W. Norton, 1987).

58. An exhaustive account of the case against grazing on the public domain is compiled by the National Public Lands Grazing Campaign (www.publiclandsranching.org) and included in G. Wuerthner and M. Matteson, *Welfare Ranching: The Subsidized Destruction of the American West* (Sausalito: Foundation for Deep Ecology, 2002).

59. According to USDA Agricultural Census (2007), five states account for the majority of beef cattle production (Texas, Kansas, Nebraska, Iowa, and Colorado). Layzer (*The Environmental Case*, 118) asserts that over 80 percent of U.S. beef pro-

duction takes place in the eastern portion of the nation upon privately owned lands. Of the livestock produced in western states, about one-third makes use of public rangelands.

60. U.S. Department of the Interior, Bureau of Land Management, *BLM and Forest Service Announce 2012 Grazing Fee*, www.blm.gov/wo/st/en/info/news room/2012/january/NR_01_31_2012.html.

61. This figure includes grazing on both national forests and BLM lands.

62. U.S. General Accounting Office (GAO-05-869), *Livestock Grazing: Federal Expenditures and Receipts Vary Depending on Agency and Purpose of Fee Charged*, September 30, 2005.

63. P. Taylor, "Obama Admin Denies Petition to Raise Grazing Fees on Public Lands," *New York Times*, January 19, 2011.

64. L. Warren, "Utah Counties Bulldoze the BLM, Park Service," *High Country News*, October 28, 1996.

65. Raymond B. Wrabley, "Managing the Monument: Cows and Conservation in Grand Staircase-Escalante National Monument," *Journal of Land, Resources and Environmental Law* 29, no. 2 (2009): 253–80.

66. Michelle Nijhuis, "Change Comes Slowly to Escalante Country," *High Country News*, April 14, 2003.

67. A more complete discussion of conservation use permits can be found on the Grand Canyon Trust web page, available at www.grandcanyontrust.org/kane/livestock-management.php.

68. See R. Glicksman and George C. Coggins, *Modern Public Land Law in a Nutshell*, 4th ed. (New York: West Publishing, 2012).

69. While collaborative resource management in the United States can be traced back to the Northwest Forest Plan crafted during the Clinton administration, community-based approaches continued to receive robust federal support through President George W. Bush's two terms and, to date, from President Obama as well.

70. The Malpai Borderlands Group has received a tremendous amount of national attention, including features in the *New York Times* and *Los Angeles Times*. The Group was singled out for special recognition by the White House, and executive director Bill McDonald received a MacArthur Foundation Genius Award in 1998. An excellent account of the Malpai Group can be found in Nathan Sayre, *Working Wilderness: The Malpai Borderlands Group and the Future of the Western Range* (Tucson: Rio Nuevo Publishers, 2005). For more on early efforts to establish collaborative and holistic approaches to rangeland management in the West, see Dan Daggett, *Beyond the Rangeland Conflict: Towards a West That Works* (Flagstaff: GCT, 1995).

71. Bill McDonald, *The Formation and History of the Malpai Borderlands Group*, 2004, www.Malpaiborderlandsgroup.org.

72. Some agreements include clauses allowing the rancher to revoke the easement if they go out of business due to circumstances beyond their control, including shifts in the market or changes in public grazing fees or stocking levels.

73. McDonald, *The Formation and History of the Malpai Borderlands Group*, 2004.

74. Ibid.

CHAPTER 8: NATIONAL WILDERNESS PRESERVATION SYSTEM: WILD AND SCENIC RIVERS AND NATIONAL SCENIC TRAILS

1. Wallace Stegner, *The Sound of Mountain Water* (Garden City: Doubleday, 1969), 147.

2. John Muir in *John of the Mountains: The Unpublished Journals of John Muir*, ed. Linnie Marsh Wolfe (Madison: University of Wisconsin Press, 1938), 313. The quote derives from Muir's journal entries penned between 1867 and 1911.

3. In the context of federal land types, the more precise phrase is *wilderness area*, which has specific legal meaning under the 1964 Wilderness Act. However, I use the word *Wilderness* here (with a capital *W*) to underscore the broader sense and meaning of the wilderness idea. It is also important to note that there may be exceptions to the list of prohibited actions depending on the specific wilderness area in question.

4. However, the same can be said of national monuments and a few other federal land classifications.

5. These values are derived from the National Wilderness Preservation System website. They are up to date as of December 27, 2012. Retrieved on January 15, 2013, at http://www.wilderness.net. This website provides exhaustive material on all units of the National Wilderness Preservation System. It is managed as a collaborative effort between the University of Montana and the federal government's wilderness training and research centers, the Arthur Carhart National Wilderness Training Center, and the Aldo Leopold Wilderness Research Institute.

6. Ibid.

7. The Federal Land Policy and Management Act of 1976 mandated the BLM to inventory its holdings for potential wilderness designations. But it wasn't until 1983 that the BLM received full managing authority for its own wilderness area, the Bear Trap Canyon, one of four units within the larger Lee Metcalf Wilderness in Montana. Prior to this, the BLM held partial authority for portions of other wilderness areas managed by the Forest Service. These included two areas designated in 1978: the Santa Lucia in California, in which the BLM managed approximately 9 percent of the 20,480-acre wilderness, and the Wild Rogue Wilderness in Oregon. In 1980, the agency also became responsible for a portion of the Frank Church-River of No Return Wilderness Area in Idaho.

8. For a full and in-depth account of the arguments pro and con for wilderness, see J. Baird Callicott and Michael P. Nelson, eds., *The Great New Wilderness Debate: An Expansive Collection of Writings Defining Wilderness from John Muir to Gary Snyder* (Athens: University of Georgia Press, 1998) and J. Baird Callicott and Michael P. Nelson, eds., *The Wilderness Debate Rages On: Continuing the Great New Wilderness Debate* (Athens: University of Georgia Press, 2008).

9. For an example, see David Brower, ed., *The Sierra Club Wilderness Handbook*, 2nd ed. (New York: Sierra Club and Ballantine Books, 1971).

10. A thorough account of this phenomenon is found in James Morton Turner, "From Woodcraft to 'Leave No Trace': Wilderness, Consumerism, and Environmentalism in Twentieth-Century America," *Environmental History* 7, no. 3 (July 2002): 462–84. See also James Morton Turner, *The Promise of Wilderness: American*

Environmental Politics Since 1964 (Seattle: University of Washington Press, 2012), 91–93. Turner provides examples of outdoor companies founded during this period, including North Face, REI, and others.

11. Cronon, "The Trouble with Wilderness," in *Uncommon Ground: Rethinking the Human Place in Nature*, ed. William Cronon (New York: W. W. Norton, 1996), 69–90. See also Callicott and Nelson, *The Great New Wilderness Debate*, 1998.

12. Henry David Thoreau, "Walking," in *The Selected Works of Thoreau*, ed. W. Harding (Boston: Houghton Mifflin, 1975), 659–86. The essay was first published a month after Thoreau's death in the June 1862 issue of the *Atlantic Monthly*.

13. An in-depth account of these events can be found in Paul Sutter, *Driven Wild: How the Fight against Automobiles Launched the Modern Wilderness Movement* (Seattle: University of Washington Press, 2002).

14. Numbers are quoted in Sutter, *Driven Wild*, 24. Population numbers listed here come from the U.S. Census Bureau, *The 2012 Statistical Abstract: Historical Statistics*, 2012. Automobile figures are drawn from the *Historical Statistics of the United States, Colonial Times to 1970* (1975), 716.

15. Established in 1919 as a citizens' watchdog group for the national park system, the National Parks Association changed its name to the National Parks and Conservation Association in 1970 to better reflect its commitment to a wide range of resource conservation issues, including water and air pollution. In 2000, the group shortened its name to the National Parks Conservation Association.

16. Robert Sterling Yard, "The Motor Tourist and the National Parks, parts 1 and 2," *National Parks Bulletin* 52 (February 1927): 11–12; 53 (July 1927): 17–19. Quoted in Sutter, *Driven Wild*, 136.

17. Carhart's official title during his years in the Forest Service was "recreation engineer."

18. Aldo Leopold, "The Wilderness and Its Place in Forest Recreation Policy," *Journal of Forestry* 19, no. 7 (1921): 718–21.

19. Frederick Jackson Turner, "The Significance of the Frontier in American History," in *Rereading Frederick Jackson Turner*, ed. John Mack Faragher (New Haven: Yale University Press, 1999), 31–60. First printed in 1893.

20. Aldo Leopold, *A Sand County Almanac, with Other Essays on Conservation from Round River* (New York: Oxford University Press and Ballantine Books, 1970). First printed posthumously in 1949.

21. Doug Scott, *The Enduring Wilderness: Protecting Our Natural Heritage through the Wilderness Act* (Golden: Fulcrum Publishing, 2004), 35.

22. Dyan Zaslowsky and T. H. Watkins, *These American Lands: Parks, Wilderness, and the Public Lands* (Washington, DC: Wilderness Society and Island Press, 1994), 201.

23. Robert Marshall, "The Problem of the Wilderness," *Scientific Monthly* 30, no. 2 (February 1930): 141–48.

24. Ibid., 148.

25. Sutter, *Driven Wild*, 227–32.

26. Ibid., 227.

27. Benton MacKaye, "An Appalachian Trail: A Project in Regional Planning," *Journal of the American Institute of Architects* 9 (October 1921): 325–30.

28. Zaslowsky and Watkins, *These American Lands*, 257. See also Sutter, *Driven Wild*, 174–5. Much of this discussion of Benton MacKaye relies on the detailed account provided by Sutter. In particular, see pages 142–93.

29. See Harvey Broome, "Origins of the Wilderness Society," *Living Wilderness* 5 (July 1940): 13–14.

30. Ibid.

31. According to author and wilderness activist Doug Scott, this had much to do with the fact that Marshall's old boss, Secretary Ickes, had given assurances of his commitment to wilderness preservation. Ickes went so far as to call for congressional action to set standards for wilderness protection in national parks. This led to a "national parks wilderness bill" introduced in Congress in 1939. The bill ultimately failed, but it helped sway support for the creation of Kings Canyon National Park in 1940. See Scott, *The Enduring Wilderness*, 33–34.

32. Zaslowsky and Watkins, *These American Lands*, 206.

33. Sutter, *Driven Wild*, 254.

34. Scott, *The Enduring Wilderness*, 35.

35. Sutter, *Driven Wild*, 255.

36. Bernard DeVoto, "Shall We Let Them Ruin Our National Parks?" *Saturday Evening Post*, July 22, 1950, 19. The piece was also published in *Reader's Digest* later that same year.

37. Scott, *The Enduring Wilderness*, 65–66.

38. Ibid., 66–73.

39. See John Miles, *Wilderness in National Parks: Playground or Preserve* (Seattle: University of Washington Press, 2009), 184–86.

40. Ibid. Additional discussions of this policy stance are found in Richard Sellars, *Preserving Nature in the National Parks: A History* (New Haven: Yale University Press, 1997), 211–13, and Turner, *The Promise of Wilderness*, 54–65. Turner provides a comprehensive account that includes not only the "Swiss cheese" approach adopted by the National Park Service but also the "purity theory" promoted by the Forest Service and the "creative approach" of the U.S. Fish and Wildlife Service.

41. Turner, *The Promise of Wilderness*, 59–60.

42. Ibid., 54–58. See also Scott, *The Enduring Wilderness*, 69–70.

43. A detailed account of this story is found in Turner, *The Promise of Wilderness*, 62–63.

44. Zaslowsky and Watkins, *These American Lands*, 215.

45. Due to a clerical error, the final form of the bill signed by President Ford contained no formal title, referred to only as Public Law 96-622. However, the bill, which initially passed through both the House and Senate, was titled the Eastern Wilderness Areas Act.

46. Endangered American Wilderness Act of 1978, PL 95-237.

47. Once RARE II was under way, the initial Roadless Area Review and Evaluation became known as RARE I.

48. The National Forest Management Act calls for new comprehensive forest management plans to be updated every ten to fifteen years.

49. Outdoor Recreation Resources Review Commission, *Outdoor Recreation for America: A Report to the President and the Congress* (Washington, DC: ORRRC, 1962).

50. The eight rivers initially protected under the Wild and Scenic Rivers Act included segments of the Feather River in California, the Clearwater and Salmon rivers in Idaho, the Eleven Point River in Missouri, the Rio Grande in New Mexico, the Rogue River in Oregon, and the Wolf and St. Croix rivers in Wisconsin.

51. Wild and Scenic Rivers Act, 82 Stat. 906, 16 USC 1271-1287, October 2, 1968.

52. National Wild and Scenic Rivers System, "Klamath River, California," http://www.rivers.gov/rivers/klamath-ca.php.

53. U.S. Department of the Interior, "Klamath Basin Water Issues," www.klamathrestoration.gov.

54. William Yardley, "Tea Party Blocks Pact to Restore a West Coast River," *New York Times*, July 18, 2012.

55. MacKaye, "An Appalachian Trail."

56. See Sutter, *Driven Wild*, 142–93.

57. See the National Trails System homepage from the National Park Service, http://www.nps.gov/nts/index.htm.

58. See the Partnership for the National Trails System, a public-private partnership designed to assist in the protection, completion, and stewardship of the national trails system, http://www.pnts.org/.

CHAPTER 9: PARTING THOUGHTS

1. John Muir, *The Writings of John Muir*, vol. 9 (Boston: Houghton Mifflin, 1916), 385.

2. For the complete argument, see Charles F. Wilkinson, *Crossing the Next Meridian: Land, Water and the Future of the West* (Washington, DC: Island Press, 1992).

3. These noncommodity perspectives refer to such notions as inherent value in nature or the important aesthetic or ecological role of various species and landscape features. In other words, they represent the values espoused in the early conservation movement.

4. The Army Corps of Engineers, which has been in continuous existence since 1802, is clearly older than the Forest Service. However, its duties are not focused primarily on public land management. Rather, the agency has a much broader mandate dealing with a wide range of military matters, including coastal fortresses and military construction sites. In the twentieth century, it became responsible for transportation on inland waterways, as well as dams for hydroelectric power generation and a variety of other services.

5. Recall that the USDA Forest Service emerged from the Division of Forestry in the Department of Agriculture before it was renamed and reorganized by Gifford Pinchot in 1905. The National Park Service came into being via congressional action in 1916 (see chapters 4 and 5 for more detail).

6. An exception to this claim relates to wilderness areas and national monuments. It is possible that large-scale wilderness area designations may continue since these tend to be drawn from lands already designated within the public land system (see chapter 8). The same holds true for national monuments. Some commentators suggest that lands held by the Department of Defense may comprise

the next opportunity for large-scale protections under the Antiquities Act, but again, these areas are already managed by federal agencies.

7. See National Park Service, "Glacier National Park: Glaciers/Glacial Features," http://www.nps.gov/glac/naturescience/glaciers.htm. The 2030 prediction is drawn from a study by M. P. Hall and D. B. Fagre, "Modeled Climate-Induced Glacier Change in Glacier National Park, 1850–2100," *Bioscience* 53, no. 2 (2003): 131–40.

8. See, for example, Karen E. Bagne and Deborah M. Finch, "Vulnerability of Species to Climate Change in the Southwest: Threatened, Endangered, and At-Risk Species at the Barry M. Goldwater Range, Arizona," *General Technical Report RMRS-GTR-284* (Fort Collins, CO: U.S. Department of Agriculture, Forest Service, Rocky Mountain Research Station, 2012): 1–139. There are many organizations that track these issues. The U.S. Fish and Wildlife Service provides information on a wide range of species threatened by climate change impacts across the United States, at http://www.fws.gov/home/climatechange/impacts.html. The Botanic Gardens Conservation International maintains an extensive list of plants around the world threatened by climate change on its website, www.bgci.org.

9. Aldo Leopold, *A Sand County Almanac, with Other Essays on Conservation from Round River* (New York: Oxford University Press and Ballantine Books, 1970), 251. First printed in 1949.

Bibliography

Abbey, Edward. *Desert Solitaire: A Season in the Wilderness*. New York: McGraw-Hill, 1968.

Adler, Jonathan H. "Stand or Deliver: Citizen Suits, Standing, and Environmental Protection." *Duke Environmental Law and Policy Forum* 12, no. 39 (2001): 39–83.

Andrews, Richard N. L. *Managing the Environment, Managing Ourselves: A History of American Environmental Policy*. New Haven: Yale University Press, 1999.

Antiquities Act. Public Law 59-209, 34 Stat. 225, 16 U.S.C. 431–33, June 8, 1906.

Babbitt, Bruce. "The Heart of the West: BLM's National Landscape Conservation System." In *From Conquest to Conservation: Our Public Lands Legacy*, 100–102. Michael P. Dombeck, Christopher A. Wood, and Jack E. Williams (Washington, DC: Island Press, 2003).

Baker, Mark, and Jonathan Kusel. *Community Forestry in the United States: Learning from the Past, Crafting the Future*. Washington, DC: Island Press, 2003.

Barringer, Mark Daniel. *Selling Yellowstone: Capitalism and the Construction of Nature*. Lawrence: University Press of Kansas, 2002.

Bean, Michael J., and Melanie J. Rowland. *The Evolution of National Wildlife Law*. Third ed. Westport: Praeger, 1997.

Behan, Richard W. *Plundered Promise: Capitalism, Politics, and the Fate of the Federal Lands*. Washington, DC: Island Press, 2001.

Brick, Philip, Donald Snow, and Sarah Van de Wetering, eds. *Across the Great Divide: Explorations in Collaborative Conservation and the American West*. Washington, DC: Island Press, 2001.

Brinkley, Douglas, and Patricia Nelson Limerick, eds. *The Western Paradox: A Conservation Reader*. New Haven: Yale University Press, 2000.

Brower, David, ed. *The Sierra Club Wilderness Handbook*. Second ed. New York: Sierra Club and Ballantine Books, 1971.

Burns, Sam. "Catron County, New Mexico: Mirroring the West, Healing the Land, Rebuilding Community." In *Forest Communities, Community Forests*, 89–115. Edited by Jonathan Kusel and Elisa Adler. Lanham: Rowman & Littlefield, 2003.

Callicott, J. Baird, and Michael P. Nelson, eds. *The Great New Wilderness Debate: An Expansive Collection of Writings Defining Wilderness from John Muir to Gary Snyder.* Athens: University of Georgia Press, 1998.

———. *The Wilderness Debate Rages On: Continuing the Great New Wilderness Debate.* Athens: University of Georgia Press, 2008.

Cawley, R. McGreggor. *Federal Land, Western Anger: The Sagebrush Rebellion and Environmental Politics.* Lawrence: University Press of Kansas, 1993.

Chaloupka, William. "The County Supremacy and Militia Movements: Federalism as an Issue on the Radical Right." *Publius* 26, no. 3 (1996): 161–75.

Clark, Ira G. *Water in New Mexico: A History of Its Management and Use.* Albuquerque: University of New Mexico Press, 1987.

Clarke, Jeanne N., and Daniel McCool. *Staking Out the Terrain: Power Differentials among Natural Resource Management Agencies.* Third ed. Albany: State University of New York Press, 1996.

Clawson, Marion. *The Federal Lands Revisited.* Washington, DC: Resources for the Future, 1983.

Clawson, Marion, and Burnell Held. *The Federal Lands: Their Use and Management.* Lincoln: University of Nebraska Press, 1957.

Clement Jr., Laurence A. "Taylor Grazing Act." In *The Encyclopedia of the Great Plains*, 55. Edited by David J. Wishart. Lincoln: University of Nebraska Press, 2011.

Coggins, George C., and R. L. Glicksman. *Public Natural Resources Law.* St. Paul: West Group, 1998.

Cooper, James Fenimore. *The Prairie.* In *The Works of James Fenimore Cooper*, Volume 2. New York: P. F. Collier, 1891.

Cronon, William. "The Trouble with Wilderness; or, Getting Back to the Wrong Nature." In *Uncommon Ground: Rethinking the Human Place in Nature*, 69–90. Edited by William Cronon. New York: W. W. Norton, 1996.

———, ed. *Uncommon Ground: Rethinking the Human Place in Nature.* New York: W. W. Norton, 1996.

Culhane, Paul J. *Public Land Politics: Interest Group Influence on the Forest Service and Bureau of Land Management.* Baltimore: Johns Hopkins University Press, 1981.

Daggett, Dan. *Beyond the Rangeland Conflict: Towards a West That Works.* Flagstaff: GCT, 1995.

Dana, Samuel T., and Sally K. Fairfax. *Forest and Range Policy.* New York: McGraw-Hill, 1980.

Dilsaver, Lary. *America's National Park System: The Critical Documents.* Lanham: Rowman & Littlefield, 1997.

Denevan, William M. "The Pristine Myth: The Landscape of the Americas in 1492." *Annals of the Association of American Geographers* 82 (1992): 369–85.

DeVoto, Bernard. "Shall We Let Them Ruin Our National Parks?" *Saturday Evening Post*, July 22, 1950, 19.

———. "The West Against Itself." In *The Western Paradox: A Conservation Reader*, 50–66. Edited by Douglas Brinkley and Patricia Nelson Limerick, 50–66. New Haven: Yale University Press, 2000. Originally published in 1947.

———. *DeVoto's West*. Columbus: Ohio State University Press, 2005.

Dombeck, Michael P., Christopher A. Wood, and Jack E. Williams. *From Conquest to Conservation: Our Public Lands Legacy*. Washington, DC: Island Press, 2003.

D'Oro, Rachel. "Alaska Residents Cash in on Annual Dividend." *Seattle Times*, September 5, 2008. http://seattletimes.com/html/localnews/2008160942_web alaskamoney05m.html.

Duncan, Dayton, and Ken Burns. *The National Parks: America's Best Idea*. New York: Knopf Doubleday, 2009.

Egan, Timothy. *The Big Burn: Teddy Roosevelt and the Fire That Saved America*. New York: Houghton Mifflin Harcourt, 2009.

Everhart, William C. *The National Park Service*. Boulder: Westview Press, 1983.

Foss, Phillip O. *Politics and Grass: The Administration of Grazing on the Public Domain*. Seattle: University of Washington Press, 1960.

Fox, Stephen. *The American Conservation Movement: John Muir and His Legacy*. Madison: University of Wisconsin Press, 1985.

Frome, Michael. *The Battle for Wilderness*. Third ed. Salt Lake City: University of Utah Press, 1997.

Gates, Paul W. *History of Public Land Law Development*. Washington, DC: GPO, 1968.

Gilpin, William. *Mission of the North American People, Geographical, Social and Political*. Philadelphia: J. B. Lippincott, 1873.

Glicksman, R., and George C. Coggins. *Modern Public Land Law in a Nutshell*. Fourth ed. New York: West Publishing, 2012.

Glover, James M. *A Wilderness Original: The Life of Bob Marshall*. Seattle: The Mountaineers, 1986.

Goldenberg, Suzanne. "The Worst of Times: Bush's Environmental Legacy Examined." *Guardian*, January 16, 2009.

Gottlieb, Robert. *Forcing the Spring: The Transformation of the American Environmental Movement*. Washington, DC: Island Press, 1993.

Haines, Aubrey L. *The Yellowstone Story*. Volumes 1 and 2, rev. ed. Boulder: University Press of Colorado, 1996 and 1999.

Harvey, Mark. *A Symbol of Wilderness: Echo Park and the American Conservation Movement*. Albuquerque: University of New Mexico Press, 1994.

———. *Wilderness Forever: Howard Zahniser and the Path to the Wilderness Act*. Seattle: University of Washington Press, 2005.

Hawthorne, Donald W. "The History of Federal and Cooperative Animal Damage Control." *Sheep and Goat Research Journal*. Paper 6. University of Nebraska-Lincoln, 2004.

Hays, Samuel P. *Conservation and the Gospel of Efficiency: The Progressive Conservation Movement*. Albuquerque: University of New Mexico Press, 1959.

———. *War in the Woods: The Rise of Ecological Forestry in America*. Pittsburgh: University of Pittsburgh Press, 2009.

———. *The American People and the National Forests: The First Century of the U.S. Forest Service*. Pittsburgh: University of Pittsburgh Press, 2009.

Hibbard, Benjamin H. *A History of the Public Land Policies*. New York: Macmillan, 1965.

Hirt, Paul W. *A Conspiracy of Optimism: Management of the National Forests Since World War Two*. Omaha: University of Nebraska Press, 1994.

Jefferson, Thomas. *Notes on the State of Virginia*. Edited by William Peden. Chapel Hill: University of North Carolina Press, 1982. First published in 1785 (France) and 1787 (United States).

Jonsson, K. "Looking Back: The Blue Goose." *USFWS Open Spaces*, January 2012. Available at http://www.fws.gov/news/blog/index.cfm/2012/1/8/Looking-Back-The-Blue-Goose.

Keiter, Robert B. *Keeping Faith with Nature: Ecosystems, Democracy and America's Public Lands*. New Haven: Yale University Press, 2003.

Kemmis, Daniel. *Community and the Politics of Place*. Norman: University of Oklahoma Press, 1990.

———. *This Sovereign Land: A New Vision for Governing the West*. Washington, DC: Island Press, 2001.

Kittredge, William. "Home on the Range." *New Republic*, December 13, 1993, 13–16.

Kline, Benjamin. *First Along the River: A Brief History of the U.S. Environmental Movement*. Fourth ed. Lanham: Rowman & Littlefield, 2011.

Klyza, Christopher M. *Who Controls Public Lands? Mining, Forestry, and Grazing Policies, 1870–1990*. Charlotte: University of North Carolina Press, 1996.

Klyza, Christopher M., and David Sousa. *American Environmental Policy, 1990–2006*. Cambridge, MA: MIT Press, 2007.

Kusel, Jonathan, and Elisa Adler, eds. *Forest Communities, Community Forests*. Lanham: Rowman & Littlefield, 2003.

Larmer, Paul. "A Bold Stroke: Clinton Takes a 1.7 Million Acre Stand in Utah." *High Country News*, September 30, 1996. http://www.hcn.org/issues/90/2795.

———. "The Mother of All Land Grabs." *High Country News*, September 30, 1996. http://www.hcn.org/issues/90/2796.

Larson/Tonopah, Erik. "Unrest in the West: Nevada's Nye County." *Time*, October 23, 1995, 52–60.

Laycock, George. *The Sign of the Flying Goose: A Guide to the National Wildlife Refuges*. Garden City: Doubleday, 1965.

Layzer, Judith A. *The Environmental Case: Translating Values into Policy*. Third ed. Washington, DC: CQ Press, 2012.

Lear, Linda. *Rachel Carson: Witness for Nature*. New York: Henry Holt, 1997.

Leopold, Aldo. "The Wilderness and Its Place in Forest Recreation Policy." *Journal of Forestry* 19, no. 7 (1921): 718–21.

———. *A Sand County Almanac, with Other Essays on Conservation from Round River*. New York: Oxford University Press and Ballantine Books, 1970. First printed in 1949.

Lewis, James G. *The Forest Service and the Greatest Good: A Centennial History*. Durham, NC: Forest History Society, 2006.

Lewis, J. "Mad-Logger." "Forgotten Characters from Forest History: Johnny Horizon." *Peeling Back the Bark*, a publication of the Forest History Society, March 17, 2011.

MacKaye, Benton. "An Appalachian Trail: A Project in Regional Planning." *Journal of the American Institute of Architects* 9 (October 1921): 325–30.

Magoc, Chris J. *Yellowstone: The Creation and Selling of an American Landscape*. Albuquerque: University of New Mexico Press, 1999.

Manning, Robert E., and Laura E. Anderson. *Managing Outdoor Recreation: Case Studies in the National Parks.* Cambridge, MA: CABI, 2012.

Marsh, George Perkins. *Man and Nature, or Physical Geography as Modified by Human Action.* New York: Charles Scribner, 1867.

Marshall, Robert. "The Problem of the Wilderness." *Scientific Monthly* 30, no. 2 (1930): 141–48.

Marston, Ed. "The Quincy Library Group: A Divisive Attempt at Peace." In *Across the Great Divide: Explorations in Collaborative Conservation and the American West,* 79–90. Edited by Philip Brick, Donald Snow, and Sarah Van de Wetering. Washington, DC: Island Press, 2001.

Merrill, Karen. *Public Lands and Political Meaning: Ranchers, the Government and the Property between Them.* Berkeley: University of California Press, 2002.

Miles, John C. *Guardians of the Parks: A History of the National Parks and Conservation Association.* Washington, DC: Taylor and Francis and NPCA, 1995.

———. *Wilderness in National Parks: Playgrounds or Preserve.* Seattle: University of Washington Press, 2009.

Muhn, James, and Hanson R. Stuart, eds. *Opportunity and Challenge: The Story of BLM.* U.S. Department of the Interior, Bureau of Land Management. Washington, DC: GPO, 1988.

Muir, John. *The Yosemite* in *The Eight Wilderness Discovery Books.* Seattle: The Mountaineers, 1992. First published in 1911.

———. *Our National Parks.* New York: Houghton Mifflin, 1901.

Nash, Roderick. *Wilderness and the American Mind,* Third ed. New Haven: Yale University Press, 1982.

National Park Service "Organic" Act. Public Law 91-458, 39 Stat. 535, 16 U.S.C. 1–4, August 25, 1916.

National Trails System Act. Public Law 90-543, 82 Stat. 919, 16 U.S.C. 1241, October 2, 1968.

Opie, John. *Nature's Nation: An Environmental History of the United States.* Fort Worth: Harcourt Brace College Publishers, 1998.

Outdoor Recreation Resources Review Commission. *Outdoor Recreation for America: A Report to the President and the Congress.* Washington, DC: ORRRC, January 1962.

Pinchot, Gifford. *The Fight for Conservation.* New York: Doubleday, Page, 1910.

Platt, Rutherford H. *Land Use and Society: Geography, Law and Public Policy.* Washington, DC: Island Press, 1996.

Prato, Tony, and Dan Fagre. *National Parks and Protected Areas: Approaches for Balancing Social, Economic and Ecological Values.* New York: Wiley-Blackwell, 2005.

Preso, Timothy. "Court Rules Continuing Supplemental Feeding on the National Elk Refuge Undermines Conservation." Earthjustice Press Release, August 4, 2011. http://www.earthjustice.org/news/press/2011//court-rules-continuing -supplemental-feeding-on-the-national-elk-refuge-undermines-conservation.

Raveling, Kristin. "Puddles Wins Twin Cities Marathon . . . Mascot Race!" *Midwest Region Field Notes.* U.S. Fish and Wildlife Service, October 6, 2012. http:// www.fws.gov/fieldnotes.

Richard, Tim, and Ellen Stein. "Kicking Dirt Together in Colorado: Community-Ecosystem Stewardship and the Ponderosa Pine Forest Partnership." In *Forest*

Communities, Community Forests, 191–206. Edited by Jonathan Kusel and Elisa Adler. Lanham: Rowman & Littlefield, 2003.

Riebsame, William E. "Ending the Range Wars?" *Environment* 38, no. 4 (1996): 4–9, 27–29.

Righter, Robert W. *The Battle over Hetch Hetchy: America's Most Controversial Dam and the Birth of Modern Environmentalism.* New York: Oxford University Press, 2006.

Robb, James, William Riebsame (Travis), and Hannah Gosnell, eds. *Atlas of the New West: Portrait of a Changing Region.* New York: W. W. Norton, 1997.

Robbins, Jim. "Suit Opposes Elk Feeding in Wyoming." *New York Times*, June 4, 2008. http://www.nytimes.com/2008/06/04/us/04refuge.html?_r=0.

Robbins, Paul. *Political Ecology: A Critical Introduction.* Second ed. New York: Wiley-Blackwell, 2011.

Roberts, John. "Plan for US Natural Gas Exports Brings Talks of Economic Boon, Fear of Failure." *Fox News*, December 17, 2012. http://www.foxnews.com/politics/2012/12/17/plan-for-us-oil-exports-brings-talk-economic-boon-fears-failure.

Rohling, Kathy, ed. *Mining Claims and Sites on Federal Lands.* P-048, BLM National Science and Technology Center. Revised May 2011.

Rohrbough, Malcolm J. *Land Office Business: The Settlement and Administration of American Public Lands, 1789–1837.* New York: Oxford University Press, 1968.

Rudzitis, Gundars. *Wilderness and the Changing American West.* New York: Wiley, 1996.

Runte, Alfred. *National Parks: The American Experience.* Fourth ed. Lanham: Rowman & Littlefield, 2010.

Sayre, Nathan. *Working Wilderness: The Malpai Borderlands Group and the Future of the Western Range.* Tucson: Rio Nuevo Publishers, 2005.

Scott, Doug. *The Enduring Wilderness: Protecting Our Natural Heritage through the Wilderness Act.* Golden: Fulcrum Publishing, 2004.

Selin, Steven W., Michael A. Shuett, and Deborah S. Carr. "Has Collaborative Planning Taken Root on the National Forests?" *Journal of Forestry* 96 (1997): 25–28.

Sellars, Richard W. *Preserving Nature in the National Parks: A History.* New Haven: Yale University Press, 1997.

Shabecoff, Phillip. "US Proposing Drilling for Oil in ANWR." *New York Times.* November 25, 1986.

Sheail, John. *Nature's Spectacle: The World's First National Parks and Protected Places.* Washington, DC: Earthscan, 2010.

Skillen, James R. *The Nation's Largest Landlord: The Bureau of Land Management in the American West.* Lawrence: University Press of Kansas, 2009.

Spence, Mark David. *Dispossessing the Wilderness: Indian Removal and the Making of the National Parks.* New York: Oxford University Press, 1999.

Steen, Harold K., ed. *The Origins of the National Forests.* Durham: Forest History Society, 1992.

——. *The U.S. Forest Service: A History, Centennial Edition.* Seattle: Forest History Society and University of Washington Press, 2004.

Stegner, Wallace. *Beyond the Hundredth Meridian.* Boston: Houghton Mifflin, 1954.

——. *The Sound of Mountain Water.* Garden City: Doubleday, 1969.

——. "Wilderness Letter." In *Marking the Sparrow's Fall: The Making of the American West*, 111–17. Edited by Page Stegner. New York: Henry Holt, 1998.

Sutter, Paul S. *Driven Wild: How the Fight against Automobiles Launched the Modern Wilderness Movement.* Seattle: University of Washington Press, 2002.

Taylor Grazing Act of 1934. Public Law 73-482, 43 U.S.C. 315-316, June 28, 1934.

Thoreau, Henry David. "Walking." In *The Selected Works of Thoreau,*659–86. Edited by W. Harding. Boston: Houghton Mifflin, 1975. First published in 1862.

Travis, William R. *New Geographies of the American West: Land Use and the Changing Patterns of Place.* Washington, DC: Island Press, 2007.

Turner, Frederick Jackson. "The Significance of the Frontier in American History." American Historical Association. Chicago World's Fair, July 12, 1893. http:// www.historians.org/about-aha-and-membership/aha-history-and-archives/ archives/the-significance-of-the-frontier-in-american-history.

Turner, James Morton. "From Woodcraft to 'Leave No Trace': Wilderness, Consumerism, and Environmentalism in Twentieth-Century America." *Environmental History* 7, no. 3 (July 2002): 462–84.

———. *The Promise of Wilderness: American Environmental Politics Since 1964.* Seattle: University of Washington Press, 2012.

Unrau, Harlan D., and G. Frank Williss. *Administrative History: Expansion of the National Park Service in the 1930s.* Washington, DC: National Park Service. http:// www.cr.nps.gov/history/online_books/unrau-williss/adhi.htm.

U.S. Department of Commerce. National Oceanic and Atmospheric Administration. *Pribilof Islands: A Historical Perspective.* http://docs.lib.noaa.gov/ noaa_documents/NOS/ORR/TM_NOS_ORR/TM_NOS-ORR_17/HTML/ Seal_Islands.htm.

U.S. Department of the Interior. Bureau of Indian Affairs. "Who We Are." http:// www.bia.gov.

———. Bureau of Land Management. *Public Land Statistics 2011.* Washington, DC: GPO, 2012.

———. *National Landscape Conservation Lands: Landscapes of the American Spirit.* Washington DC: GPO, 2012.

———. National Park Service. "Yellowstone Bison." http://www.nps.gov/yell/ naturescience/bison.htm.

———. National Park Service. "National Park System Overview." http://www .nps.gov/news/upload/NPS-Overview-2012_updated-04-02-2013.pdf.

———. U.S. Fish and Wildlife Service. "History of the National Wildlife Refuge System." http://www.fws.gov/refuges/history.

———. *The National Elk Refuge: A Legacy of Conservation.* Washington, DC: GPO, 2008.

———. *Annual Report of Lands under Control of the U.S. Fish and Wildlife Service* (Washington, DC: GPO, 2012).

———. "Everglades Headwater National Wildlife Refuge and Conservation Area." http://www.fws.gov/southeast/evergladesheadwaters.

———. "Legislative Mandates and Authorities for the U.S. Fish and Wildlife Service." http://www.fws.gov.

Weber, Edward P. *Bringing Society Back In: Grassroots Ecosystem Management, Accountability and Sustainable Communities.* Cambridge, MA: MIT Press, 2003.

Wheeler, David L. "The Blizzard of 1886 and Its Effect on the Range Cattle Industry in the Southern Plains." *Southwestern Historical Quarterly* 94 (1991): 415–32.

White, Richard. *"It's Your Misfortune and None of My Own": A New History of the American West.* Norman: University of Oklahoma Press, 1991.

Wild and Scenic Rivers Act. Public Law 90-542, 82 Stat. 906, 16 U.S.C. 1271-1287. October 2, 1968.

Wilderness Act. Public Law 88-577, 78 Stat. 890, 16 U.S.C. 1131. September 3, 1964.

Wilkinson, Charles F. *Crossing the Next Meridian: Land, Water and the Future of the West.* Washington, DC: Island Press, 1992.

Williams, Gerald W. *The Forest Service: Fighting for Public Lands.* Westport: Greenwood Press, 2007.

Wilson, Paul. "The Public Trust Doctrine in Wildlife." *Mountain State Sierran Newsletter.* Sierra Club, 2005.

Wilson, Randall K. "'Placing Nature': The Politics of Collaboration and Representation in the Struggle for La Sierra in San Luis, Colorado." *Ecumene* 6 (1999): 1–28.

———. "Community-Based Management and National Forests in the Western United States: Five Challenges." *Policy Matters* 12 (September 2003): 216–24.

———. Narratives of Community-Based Resource Management in the American West." In *Rural Change and Sustainability: Agriculture, the Environment, and Communities,* 343–57. Edited by Stephen Essex, Andrew Gilg, Richard Yarwood, John Smithers, and Randall K. Wilson. Wallingsford, UK: CABI Publishing, 2005.

———. "Collaboration in Context: Rural Change and Community Forestry in the Four Corners." *Society and Natural Resources* 19 (2006): 53–70.

———. "Tracking Collaboration: Forest Planning and Local Participation on the San Juan National Forest, Colorado." *Local Environment* 13 (2008): 609–25.

Wolfe, Linne Marsh, ed. *John of the Mountains: The Unpublished Journals of John Muir.* Second ed. Madison: University of Wisconsin Press, 1938.

Wondolleck, Julia M., and Steven L. Yaffee. *Making Collaboration Work: Lessons from Innovation in Natural Resource Management.* Washington, DC: Island Press, 2000.

Worster, Donald. *The Dust Bowl: The Southern Plains in the 1930s.* New York: Oxford University Press, 1979.

———. *Rivers of Empire: Water, Aridity and the Growth of the American West.* New York: Pantheon Books, 1985.

———. *A Passion for Nature: The Life of John Muir.* New York: Oxford University Press, 2008.

Wrabley, Raymond B. "Managing the Monument: Cows and Conservation in Grand Staircase-Escalante National Monument." *Journal of Land, Resources and Environmental Law* 29, no. 2 (2009): 253–80.

Wyckoff, William, and Larry M. Dilsaver. *The Mountainous West: Explorations in Historical Geography.* Lincoln: University of Nebraska Press, 1995.

———. "Promotional Imagery of Glacier National Park." *Geographical Review* 87 (January 1997): 1–26.

Yard, Robert Sterling. "The Motor Tourist and the National Parks, parts 1 and 2." *National Parks Bulletin* 52 (February 1927): 11–12; 53 (July 1927): 17–19.

Yardley, William. "Tea Party Blocks Pact to Restore a West Coast River." *New York Times,* July 18, 2012.

Yochim, Michael J. "The Development of Snowmobile Policy in Yellowstone National Park." *Yellowstone Science* (Spring 1999): 2–10.

Zaslowsky, Dyan, and T. H. Watkins. *These American Lands: Parks, Wilderness, and the Public Lands.* Washington, DC: Wilderness Society and Island Press, 1994.

Index

About the Author

Randall K. Wilson is professor of environmental studies at Gettysburg College in Pennsylvania, where he teaches courses on environmental policy, natural resource management, sustainable communities, and the geography of the American West. He earned degrees from Humboldt State University in California and the University of Colorado, before moving to the University of Iowa to study for a doctorate in geography. Professor Wilson is coeditor of *Rural Change and Sustainability: Agriculture, the Environment and Communities* and has conducted research and published numerous works pertaining to collaborative conservation on U.S. public lands, rural transformation, and the political ecology of the American West. He recently served as a Fulbright Scholar in Austria at the University of Vienna.